COSTS AND BENEFITS OF RESERVE PARTICIPATION

New Evidence from the 1992

Reserve Components Survey

Sheila Nataraj Kirby ◆ **David Grissmer**

Stephanie Williamson ◆ **Scott Naftel**

Prepared for the
Office of the Secretary of Defense

National Defense Research Institute

RAND

Operation Desert Shield/Storm (ODS/S) was the first major mobilization of reserve forces in almost 50 years, and the first since the advent of the All-Volunteer Force. Since then, ODS/S reservists have participated in every overseas military operation including those in Somalia, Haiti, and Bosnia. This participation indicates the greater reliance on reserve forces in the post-drawdown environment. Reservists will likely be used in every future military action as well as in overseas operations that have traditionally been exclusively the province of the active force.

This greater reliance on reserve forces makes it important to determine how reserve mobilizations affect the attitudes, perceptions, and behaviors of reservists, their families, and their employers. It is possible that their greater use will significantly change important attitudes and behaviors and will require changed personnel and training policies in response. This study examines whether key attitudes and perceptions of reservists and the level of support they receive from their civilian employers and families have changed in significant ways. We do this in two ways: first, by comparing survey data collected from reservists in 1986 and 1992 and second, by comparing the responses of mobilized and nonmobilized reservists who responded to the 1992 survey. This information could be crucial in determining whether new or stronger policies are needed to protect reservists during periods of mobilization and in foreshadowing problems that might be associated with a policy of increased reserve use.

This work was sponsored by the Assistant Secretary for Reserve Affairs. The research was conducted in the Forces and Resources Policy Center, which is part of RAND's National Defense Research Institute, a federally funded research and development center sponsored by the Office of the Secretary of Defense, the Joint Staff, and the defense agencies.

CONTENTS

TABLES

BACKGROUND

Before Operation Desert Shield/Storm (ODS/S), the last major mobilization of reservists occurred almost 50 years ago during the Korean conflict. Thus, reservists serving in the Selected Reserve Components before 1990 had largely no experience with a large-scale reserve mobilization. In fact, the likelihood of a reserve mobilization was probably viewed as so remote that it played almost no role in decisions to join or remain in the reserve forces. The ODS/S mobilization, and the subsequent involvement of reservists in various other operations, ranging from the Army Multinational Force and Observers (MFO) Sinai Initiative to the current effort in Bosnia, have changed this perception in important ways.

It is clear that the Reserve Components are expected to play an important role in responding to regional crises, as well as in peacekeeping, peace enforcement, and humanitarian assistance operations. For instance, since 1991, Reserve Component members were activated or volunteered to support Operation Restore Democracy (Haiti), Provide Promise and Deny Flight (Bosnia), Restore Hope (Somalia), Southern Watch (Southern Iraq), and Provide Comfort (Northern Iraq). Mobilizations are likely to be more frequent in the future and are likely to have important effects on reservists' attitudes and the degree of support they receive from their families and civilian employers.

Understanding how mobilizations affect reservists is important for three reasons: (a) The increasing reliance on reserves means increased chances of mobilization for reservists, (b) the lack of mobilization experience before ODS/S means that there is little empirical research concerning the effects mobilizations have on reservists' attitudes and those of their employers and families, (c) reservists' decisions to stay in the reserve are critically dependent on their own attitudes and perceptions and those of employers and families. If mobilizations change these attitudes in significant ways, then retention and recruiting in the years after mobilization might be affected with potentially important effects on reserve personnel readiness. If each mobilization leaves lasting imprints on reserve personnel and their decisions to enlist and stay in the reserves, this may lead to a gradual but steady reshaping of the force with unforeseen and perhaps unwanted consequences.

DATA AND PURPOSE OF THE REPORT

The analyses reported here are based on a comparison of two large-scale surveys. The 1986 and 1992 Reserve Components Surveys are two in a series of periodic surveys of officers and enlisted personnel conducted by the Department of Defense (DoD) to collect information regarding the morale, perceptions, and civilian characteristics of reservists. Results from the 1986 survey reported in Grissmer, Buddin, and Kirby (1989) are compared to similar analyses of the 1992 survey. This allows us to examine changes in key variables over time.

The second focus of the analysis is a comparison of mobilized and nonmobilized reservists in 1992. The 1992 survey is a rich source of experiential information on the attitudes and problems faced by mobilized reservists, allowing us to determine more directly the effect of a large mobilization such as ODS/S on those who were mobilized. We examine differences between mobilized and nonmobilized reservists not only in terms of their perceptions and attitudes about the reserve, their families, and their work environments, but also their rankings of the potential problems they would face if called up. What makes the latter particularly interesting is that for one group, these rankings are based on experience; for the other, they are based on judgment.

It is important to be clear about what this report does and does not do. It is a simple comparative analysis of the changes between 1986 and 1992 in attitudes and perceptions of reservists regarding their reserve participation, unit readiness, and family and work environments. We hypothesize that these changes, if any, are likely to have been caused primarily by ODS/S and perhaps to a lesser extent by the drawdown attendant on the end of the Cold War, but we do not explicitly test these hypotheses in a multivariate framework. For policy purposes, it is useful to measure changing attitudes and behavior. These results can help determine the need for changes in personnel and mobilization policies and anticipate (and forestall) problems likely to arise in future mobilizations.

FINDINGS

From FY86 to FY92, the Selected Reserve has become increasingly senior and more experienced. In addition, the quality of the force has improved significantly. Despite this, the attitudes, characteristics, and family and work environments of reservists in 1992 are remarkably similar to those reported by reservists in the 1986 survey.[1]

RETENTION-RELATED ATTITUDES AND PERCEPTIONS

Motivation for Staying in the Guard/Reserve, 1986 and 1992

The motivation for staying in the guard/reserve appears to have shifted from 1986 to 1992.

- Among enlisted personnel, there is less emphasis on immediate compensation and promotion and greater importance placed on educational benefits; among officers, patriotic and job satisfaction motives were more frequently mentioned. Expanded educational benefits may have attracted a newer group of young en-

---

[1]We should point out that the 1992 survey has a 50 percent nonresponse rate and that young, single, black, junior reservists tended to have the highest rates of nonresponse. If these reservists have very different attitudes/perceptions/behaviors than their counterparts, then the nonresponse weighting adjustment and subsequent poststratification will not fully compensate for the extent of nonresponse bias (see Appendix A for further details). However, much of our analysis excludes the most junior paygrades (E-1 and E-2); to some extent, we have avoided the problem of drawing inferences about groups that have very low response rates. This does not fully address the more general issue that plagues all survey data—the extent of the nonresponse bias that occurs when nonrespondents are not a random subset of the total sample in each subgroup.

listed personnel whose primary motivation is obtaining money for college rather than long-term reserve service.

- There is, however, a small but definite increase in the levels of dissatisfaction with military pay and opportunities for education/training among both officers and enlisted personnel. Part of the dissatisfaction with pay may be a reflection of the perceived higher risk of mobilization and the attendant likely economic losses.

Civilian Work Environment

- Perhaps the most important positive change is the shift in employer attitudes from 1986 to 1992: Reservists report a more favorable attitude on the part of their civilian supervisors in 1992 than in 1986. This may have been partly a result of the significant contributions reservists made during ODS/S.

- This shift seems to have lessened the conflicts reservists traditionally feel between fulfilling reserve obligations and those of their civilian job. Reservists report much less conflict with employers about attending drills and annual training and in spending extra time on reserve obligations while on the job. (In the latter case, this may be due to fewer demands from the reserve job for extra time.)

Family Attitudes/Support

- Perceived attitudes of spouses have remained stable between 1986 and 1992 (a little surprising, given the increased chance of mobilization). These data are particularly important because of the importance of spouses' attitudes in reenlistment/continuation decisions (Grissmer, Kirby, and Sze, 1992).

- There also appears to be about the same or less conflict with family time arising from reserve drills, annual training, and extra time spent on reserve obligations.

Mobilized and Nonmobilized Reservists

Comparing mobilized with nonmobilized reservists, we find significant differences in the reported attitudes of spouses and civilian supervisors.

- Higher proportions of mobilized officers reported unfavorable attitudes on the part of both spouses and civilian supervisors compared to nonmobilized reservists.

- Among enlisted personnel, we find an increased incidence of unfavorable spouse attitudes among mobilized reservists, but little or no difference in supervisor attitudes.

- Where differences exist, they tend to be much larger among the junior ranks.

- In addition, junior mobilized officers and enlisted personnel were also much more dissatisfied with pay and benefits than nonmobilized personnel.

- However, overall satisfaction with reserve service showed little difference between mobilized and nonmobilized personnel.

Continuation Rates

This report cannot speak definitively about the effect of mobilization on retention; that is the subject of another report. Nonetheless our data suggest the following:

- Although reservists across most grades reported much lower subjective probabilities of reenlistment/continuation in the 1992 survey than in the 1986 survey, our simple analysis of continuation rates found little difference between 1986 and 1992. Our data show no dramatic change in overall behavior that could be attributable to ODS/S. However, we must caution that this finding is based on very simple, aggregated measures.

- There is little or no difference in the overall retention rates of mobilized and nonmobilized reservists; however, among officers who expressed serious doubts about continuing, mobilized reservists had much lower retention rates than nonmobilized officers.

PERCEIVED PROBLEMS IN MEETING UNIT TRAINING OBJECTIVES, 1986 AND 1992

It is clear that the majority of reservists do not perceive serious problems in their unit's ability to meet training objectives. Even problems that ranked in the top five were mentioned by only 10–16 percent of the enlisted and by only 10–20 percent of officers.

- There is remarkable similarity in the 1986 and 1992 groups in the level and type of concern expressed by reservists about the problems facing units in meeting training objectives. If we omit uncertainty about the future status of the unit (not an issue in 1986), we find that the lack of time for planning and administration, lack of access to good training facilities, and lack of supplies and modern equipment/weapons remain the primary concerns in both surveys.

- There is little difference in the rankings of problems by mobilized and nonmobilized reservists, suggesting that the experience of mobilization has not changed perceptions.

- There is a fair amount of consistency in the problems mentioned by the reservists in the different components, although there is a difference in their perceptions of how serious these problems are. By and large, the air components seem pretty satisfied with their ability to meet training objectives but the naval reserve and the two army components are somewhat less optimistic.

POTENTIAL PROBLEMS FOR FUTURE RESERVE MOBILIZATIONS

As Figures S.1 and S.2 show, family and economic issues dominate the list of problems that reservists could potentially face if mobilized.

- Potential loss of income is the most important concern of reservists, mentioned by 35–40 percent of the reservists;

- Burden on spouses and increased family problems are mentioned by 20–30 percent;

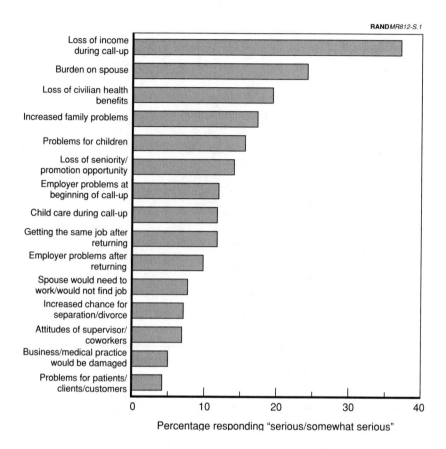

RAND*MR812-S.1*

Percentage responding "serious/somewhat serious"

Figure S.1—Potential Problems for Enlisted Personnel If Called Up, 1992

- The loss of civilian health benefits ranks third, mentioned by one-fifth of the reservists.

- Employer-related concerns—problems with employers when mobilized and returning, getting the same job back, damage to business practice, problems for clients and patients—are somewhat lower down on the list.

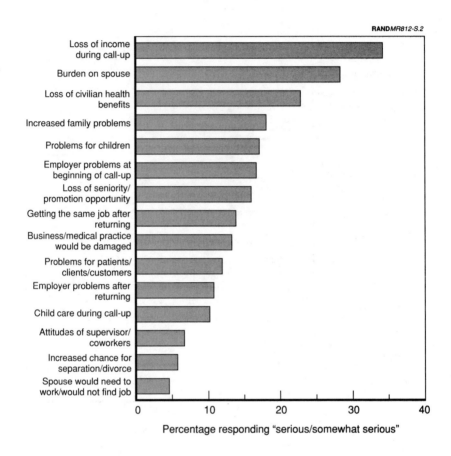

Figure S.2—Potential Problems for Officers If Called Up, 1992

- Potential problems during mobilization vary markedly among different groups of reservists: Self-employed reservists and doctors naturally express very high levels of concern regarding income loss and damage to business or practice; loss of civilian health benefits ranks much higher among pilots than any other group; family problems (burden on spouse, problems for children, etc.) weigh heavily on the minds of those with families. It is important to be aware of these differences when mobilizing specific groups of reservists.

Mobilized and Nonmobilized Reservists

Figures S.3 and S.4 rank the problems reported by mobilized and nonmobilized enlisted personnel and officers. As the data make clear, mobilized and nonmobilized reservists do not have drastically different perceptions of problems in a future mobilization, suggesting that reservists have relatively accurate perceptions of the problems they are likely to face if called up, with a couple of exceptions. Nonmobilized reservists are more concerned about income loss and loss of civilian health care benefits than are mobilized reservists,

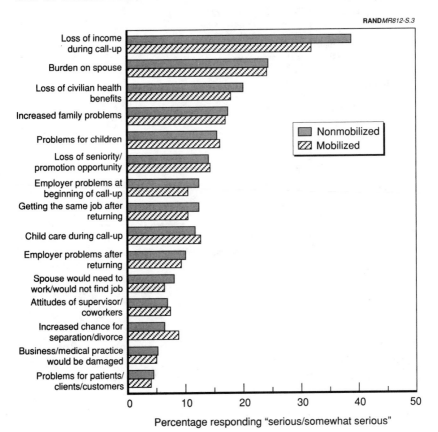

Figure S.3—Potential Problems for Reservists If Called Up: Mobilized and Nonmobilized Enlisted Personnel, 1992

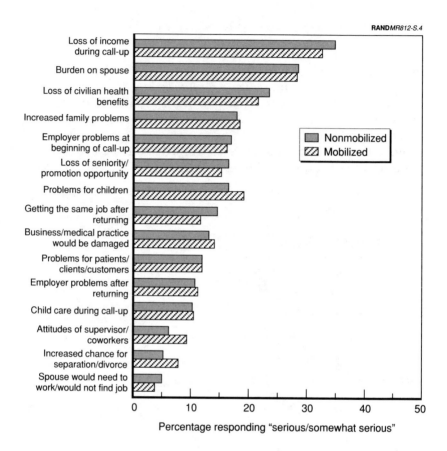

RAND*MR812-S.4*

Figure S.4—Potential Problems for Reservists If Called Up: Mobilized and
Nonmobilized Officers, 1992

suggesting that the reality was not as bad as expected; on the other
hand, mobilized reservists express more concern about marital stability and concern for children in future mobilizations.

CAVEATS

This report is a simple descriptive analysis of changes in the attitudes, perceptions, and family and work environments of reservists
in 1986 and 1992. Part of our focus in this report was on differences

between mobilized and nonmobilized reservists in 1992; this is an important addition, given that these data provide a first look at how mobilization affects family life, work environments, and the attitudes and behavior of those who were called up.

However, there are two main caveats that must be kept in mind:

- Our data allow us to make inferences regarding the effect of ODS/S on reservists, but we cannot generalize our findings to future mobilizations that might be very different from ODS/S in terms of magnitude, duration, or popularity. We cannot afford to become complacent about the likely effects of future mobilizations on the reserves. The many operations in which reservists have been (and continue to be) used since ODS/S need to be carefully studied to gauge the likely effects of different types of mobilizations.[2]

- It is difficult to assess the effect of increased use of reservists on reserve retention a priori. The increased chance to contribute in meaningful ways to real operations may need to be balanced against the increased likelihood of conflicts with both employers and families. In addition, as the nature and terms of the reserve contract are seen to change, it is difficult to predict the effect on recruiting for similar reasons. Given the importance of the reserves in our military strategy, it is essential that manpower planners continue to monitor these indicators and to be proactive in forestalling problems in both areas.

POLICY IMPLICATIONS

The chief concern of reservists is potential economic losses if they were mobilized. The Department has considered several options to address this issue including a form of mobilization insurance.

Another major concern is the loss of civilian health care benefits. Although reserve families become eligible for military health care upon mobilization, the loss of civilian benefits can cause serious problems in the continuity and transaction costs associated with medical care.

_____

[2]There is a study currently under way at RAND looking at this question.

For those losing civilian benefits—a large group, consisting of almost half of those who had such a benefit before mobilization—the burden remains on the family to work out issues of location and access, possible transfer of medical records, and obtaining continuing care for chronic conditions. It would be in the interest of families and probably the reserve forces to direct efforts at maintaining the same civilian-provided health care arrangements for families when reservists are mobilized. This might be done through special CHAMPUS[3] reimbursement mechanisms or through direct payments to families or employers for maintaining civilian health care benefits. However, policies would have to be shaped so that employers who maintain coverage do not shift responsibility to the government.

Further focus on family-oriented programs for mobilized reservists directed at support of spouses and children seems warranted. The precipitous shift of burdens and responsibilities to spouses upon mobilization, the related effects on children, and possible risk to marital stability might be alleviated through improved access to counseling and support during mobilization, especially during the transition periods, which are likely to be particularly stressful.

Educational benefits were a much more important consideration for younger enlisted personnel in 1992 than in 1986 and may prove an important drawing card for the reserves in the future. These benefits appear to attract higher-quality enlistees who use college benefits while serving in the reserve; in general, we find that higher-quality recruits appear to stay longer in the reserve (Grissmer and Kirby, 1988; Kirby and Grissmer, 1993; Buddin and Kirby, 1997). However, the question of whether recruits who entered primarily to obtain the educational benefits rather than because of a desire to serve in the reserve are likely to have lower retention is an important, although largely unanswered, one.

---

[3]Civilian Health and Medical Program of the Uniformed Services.

ACKNOWLEDGMENTS

We are grateful to Wayne Spruell, Daniel Kohner, and COL. Richard Krimmer for their support of this study and the Defense Manpower Data Center for providing the data that formed the basis of the analysis. Two RAND reviewers, Albert Robbert and Narayan Sastry, provided very thoughtful comments. The report benefited greatly from their constructive criticism, both substantively and in terms of increased clarity and organization. We also thank Cathy Montalvo and Birthe Wenzel for their able secretarial assistance with the report, and Patricia Bedrosian for her impeccable editing.

ACRONYMS

AFR	Air Force Reserve
AGRs	Active Guard/Reservists
ANG	Air National Guard
ARNG	Army National Guard
CHAMPUS	Civilian Health and Medical Program of the Uniformed Services
DoD	Department of Defense
ETS	End of Term of Service
IADT	Initial Active Duty Training
IMAs	Individual Mobilization Augmentees
MCR	Marine Corps Reserve
MFO	Multinational Force and Observers
NR	Naval Reserve
ODS/S	Operation Desert Shield/Storm
USAR	United States Army Reserve

Chapter One

INTRODUCTION

Although the Total Force Policy placed more reliance on reserve forces, it was not until Operation Desert Shield/Storm (ODS/S) that the reality of that reliance was made apparent. The drawdown of active forces since ODS/S has shifted even more reliance onto reserve forces.[1] This reliance is evidenced by reserve participation in every foreign deployment since ODS/S. For instance, since 1994, Reserve Component members were activated or volunteered to support Operation Restore Democracy (Haiti), Provide Promise and Deny Flight (Bosnia), Restore Hope (Somalia), Southern Watch (Southern Iraq), and Provide Comfort (Northern Iraq).[2]

---

[1]On April 7, 1995, the Secretary of Defense, William Perry, in a memorandum to the services and the Joint Staff, outlined the need for increased use of reserve forces in Total Force missions:

> As the Armed Forces of the U.S. are being drawn down in accordance with our National Security Strategy, we continue to ask the Active components to meet increasingly demanding operational requirements. We need to better leverage our National Guard and Reserve forces, which are well qualified and capable of performing some of these missions. In the Cold War, the emphasis for the Active components was on fulfilling operational requirements, and the focus for the Reserve components was on training for mobilization. We need to reorient our thinking and plan to capitalize on Reserve component capabilities to accomplish operational requirements while maintaining their mission readiness for overseas and domestic operations . . . .

> Increased reliance on the Reserve components is prudent and necessary in future policy, planning, and budget decisions.

[2]For example, the Army has used reserve chaplains, contract and environmental law specialists, automation specialists, biotechnology experts, pathologists, and marketing and media relations experts on a variety of occasions; the Navy has used reservists for dental support at active facilities, telecommunications support, scientific and techni-

Before ODS/S, the last major mobilization of reservists occurred almost 50 years ago during the Korean conflict. Since then, there have been only very infrequent call-ups involving small numbers of reservists for short periods. The majority of personnel serving in the reserve forces before ODS/S had virtually no experience with a reserve mobilization. The probability of being mobilized was probably viewed as such an unlikely and remote possibility that it played almost no role in decisions to join or remain in the reserve forces. The ODS/S mobilization, where more than one in five reservists were mobilized, definitely changed this perception and potentially has affected the decision calculus.

Understanding how mobilizations of reservists affect their attitudes and those of their families and employers is important for at least two reasons. First, defense policy in the post-Cold War, post-drawdown era has increased reliance on reserve forces, and practically any overseas involvement—from small scale to large scale—will involve reserve forces. Since there was little experience before ODS/S in mobilizing reservists, there is little empirical research concerning the effects mobilizations have on reservists' attitudes and those of their employers and families. Second, as our earlier research has shown (Grissmer, Kirby, and Sze, 1992), reservists' decisions to stay in the reserve are critically dependent on their own attitudes and perceptions and those of employers and families. If mobilization— or the increased likelihood of being mobilized—changes these attitudes, then retention patterns may change and this will have a significant effect on the type and characteristics of people in the reserve. In addition, willingness to join the reserve may also be affected, leading to a gradual reshaping of the force, perhaps in ways unforeseen by manpower planners.

Changed attitudes and perceptions could also affect reservists' choices of occupations and units, their job performance, and eventually the readiness of units. For instance, earlier work (Grissmer, Kirby, and Sze, 1995) has documented extensive economic losses for mobilized reservists as a result of ODS/S, particularly for specific groups such as health care personnel; this may lead to future shortages in these critical areas.

---

cal research, and legal assistance; the Air Force employs reservists to provide intelligence, medical, legal, engineer, and public affairs support.

Tracking and comparing attitudes and perceptions of mobilized and nonmobilized reservists may prove helpful in preparing for future mobilizations, ensuring unit readiness, and forestalling potential problems. For example, individuals who were mobilized may bring back different attitudes toward their own unit's readiness, and a different set of perceptions about what factors are important in maintaining readiness. Although these data represent reservists' *perceptions* about unit readiness, rather than objective measures of actual unit readiness, they can, nonetheless, provide useful information about the confidence reservists have in their unit's ability to perform its assigned mission and what they perceive to be lacking or potential problems.

It should be emphasized that the effects of mobilization can potentially be either positive or negative. The experience of being mobilized may well enhance unit cohesion and foster a sense of pride and belonging that may prove invaluable in readiness. A popular mobilization, such as ODS/S, may increase family and employer support and so increase retention; however, this effect is likely to be heavily dependent on the circumstances, length, and frequency of mobilization. So our results for ODS/S may not be generalizable to conflicts and mobilizations with different characteristics and it is important to be aware of the limitations of the analysis.

This report is an initial empirical investigation of whether and how the perceptions of reservists have changed over time and whether the experience of mobilization affected reservists' own attitudes and home and work environments in important ways.

It is important to be clear about what this report does and does not do. It is a simple comparative analysis of the changes between 1986 and 1992 in attitudes and perceptions of reservists regarding their reserve participation, unit readiness, and family and work environments. We hypothesize that any changes are likely to have been caused primarily by ODS/S and perhaps to a lesser extent by the drawdown attendant on the end of the Cold War, but we do not explicitly test these hypotheses in a multivariate framework. The purpose of this report is to determine if attitudes have changed between 1986 and 1992 and whether there are differences between mobilized and nonmobilized reservists to determine the need for and feasibility of a more detailed, more analytical report. For policy purposes, it is useful to measure changing attitudes regardless of causes.

In addition, we look at potential problems facing reservists during mobilizations, as reported in the survey. The experiential data gathered from mobilized reservists are likely to prove a rich source of information regarding such problems and their relative importance.

These results can help policymakers establish improved personnel and mobilization policies based on the experience of ODS/S and anticipate problems and changing attitudes resulting from future mobilization. Such information could be crucial in determining whether new or stronger policies are needed to protect reservists (from income losses, loss of benefits, employer and family problems) during periods of mobilization and in foreshadowing potential problems that might be associated with a policy of increased reserve use.

DATA

The 1986 Reserve Components Survey is one in a series of periodic surveys of officers and enlisted personnel conducted by the Department of Defense (DoD) to collect information on the morale, perceptions, and civilian characteristics of reservists. It was fielded in the spring of 1986 and the survey population consisted of officer and enlisted personnel who were attending drills, thus excluding non-prior-service personnel at Initial Active Duty Training (IADT). The basic stratification variable was the Reserve Component. Within each component, personnel were classified by reserve category (unit members, Individual Mobilization Augmentees (IMAs), military technicians, and full-time support personnel). In most strata, the survey design provided for a 10 percent sample. The adjusted response rate was 60 percent for enlisted personnel and 76 percent for officers. However, for this report, we did not reanalyze the 1986 data but depended instead on published data reported in Grissmer, Buddin, and Kirby (1989). The 1989 analysis was limited to part-time reservists, which included unit members and IMAs.

The 1992 DoD Reserve Components Survey, which forms the main focus of this report, is another in the series of periodic surveys of reserve personnel. The 1992 survey sample was drawn from the December 1991 reserve population and was updated with current addresses and pay grade as of March 1992.[3] The survey was in the field from October 1992 through late 1993; as a result, the eligible population was redefined to include only those reservists who were in the reserve in both December 1991 and October 1992. The sampling rate was 0.08 of the total population, which resulted in an eligible sample size of a little over 72,000. The overall (adjusted) response rate was 50 percent. For details of the sample design, the weighting procedure, and nonresponse, see Appendix A. To be consistent with the 1989 report, we limited the analysis to part-time reservists only (again, unit members and IMAs);[4] further, for some of the analyses, we omitted E-1s and E-2s and O-7s and O-8s, as was done with the 1986 analyses reported in Grissmer, Buddin, and Kirby (1989). We excluded the former because they are generally too new to the reserve to provide credible or thoughtful answers to many of the survey questions; in addition, they had by far the largest nonresponse rates. The senior officers were excluded because of their very small sample size (an unweighted count of 63 in the 1992 final sample). The final respondent sample size was around 32,000 reservists.

In addition, reservists were asked whether they were mobilized for ODS/S. This allowed us to categorize reservists as mobilized or nonmobilized and to examine differences between them not only in their perceptions and attitudes about the reserve, family, and work environment but also in their rankings of potential problems they

---

[3]The sample excluded those who either left or entered the reserve between December 1991 and March 1992.

[4]The 1992 reserve sample consists of both a longitudinal component and a members' component. The longitudinal component is drawn from the sample of reservists sampled in 1986; the members' sample is a stratified random sample drawn from the 1992 frame *after* excluding the 1986 sample reservists. Rizzo et al. (1995) point out that the design can be viewed in two ways: (a) The 1986 sample group can be viewed as a fixed portion of the 1992 population, in which case "the two components are samples from two mutually exclusive and exhaustive strata" (p. B-9); or (b) it can be viewed as a double sample process in which we draw a sample in 1986 from all 1986 reservists and then draw a second sample from the 1986 sample members still in the Selected Reserve in 1992 for the longitudinal component; the members' component is then drawn from the remainder of the 1992 reserve population. In any case, it seems clear that we need to include both the longitudinal component and the unit member samples to draw inferences about the 1992 reserve population. We use the full sample for the analyses presented here because we were interested in more than just comparisons between the 1986 and 1992 samples.

would face if called up. What makes the latter particularly interesting is that for one group, these rankings are based on experience, whereas for the other, they are based on judgment.

ORGANIZATION OF THE REPORT

The second chapter provides a brief profile of the Selected Reserve in 1986 and 1992. The third chapter compares the attitudes of reservists, their families, and their employers in 1986 and 1992 on a variety of measures previously shown to be related to retention behavor. The focus here is on the costs and benefits of reserve participation and whether and how these may have changed from 1986 to 1992. The fourth chapter analyzes differences between mobilized and nonmobilized reservists along the same dimensions.

The focus thus far is on individual circumstances. The fifth and sixth chapters examine unit readiness: (a) whether perceptions of unit readiness have changed from 1986 to 1992 and in what direction; and (b) whether mobilized and nonmobilized reservists rate their units differently.

The seventh chapter presents perceptions of problems for future mobilizations and differences in these perceptions for mobilized and nonmobilized reservists. Chapter Eight presents conclusions.

Chapter Two

A PROFILE OF THE SELECTED RESERVE:
FY86 AND FY92

BACKGROUND

The main purpose of this report is to compare the environment facing Selected Reservists in 1986 and 1992. The major change that occurred during this time period that could explain the attitudes and perceptions of reservists in this period is Operation Desert Shield/Storm (ODS/S).[1] ODS/S was the first large-scale mobilization of reservists since the Korean war. Approximately 250,000 of the 1.1 million reservists were mobilized for ODS/S and most served for 5–6 months between August 1990 and March 1991.

However, before we can reasonably attribute observed changes in attitudes and perceptions to the ODS/S experience, we need to make sure that the characteristics of reservists remained stable from 1986 to 1992. If they did not, then some of these changed characteristics might well account for the changes in attitudes.

A PROFILE OF THE SELECTED RESERVE, FY86 AND FY92

The Selected Reserve forces have grown somewhat larger since 1986. By the end of FY92, the endstrength stood at about 1.1 million, an

---

[1]The end of the Cold War occurred in this period also. However, the downsizing of the forces as a result of this did not start until FY94—and Selected Reserve downsizing was less than for active forces. Yet, it is possible that the anticipated changes in force sizes and possible downsizing were present as early as 1991/1992.

increase of approximately 5 percent since 1986. However, there has been no dramatic shift in each component's share of the force (Figure 2.1).

The FY92 force is more experienced than the FY86 force, as Figures 2.2 and 2.3 show. The proportion of nonprior-service recruits (approximated by those with 0–2 years of service) is considerably lower in FY92 than in FY86 whereas the proportion of those with 20 years of service or more is considerably higher. High retention rates among reserve personnel are the primary reason; this reduces turnover and the demand for new nonprior-service accessions. This is true among both enlisted personnel and officers.

Figures 2.4 and 2.5 present the rank distribution of enlisted personnel and officers for the two years. Reflecting the greater experience of the force in FY92, we find a much larger proportion of mid-career and senior enlisted people (E-4 through E-8) in the FY92 force compared with the FY86 force, with corresponding reductions in the proportions of junior paygrades. This trend reflects higher retention, lower levels of nonprior-service accessions, and somewhat higher levels of prior-service accessions who enter the reserve after

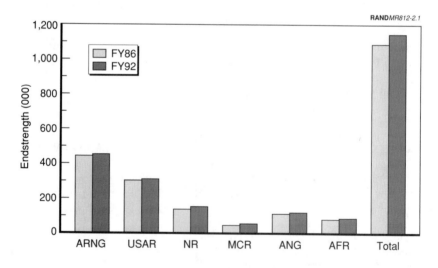

Figure 2.1—Selected Reserve Endstrength by Component,
FY86 and FY92

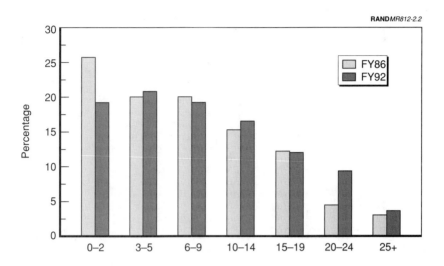

Figure 2.2—Distribution of Enlisted Personnel by Years of Service,
FY86 and FY92

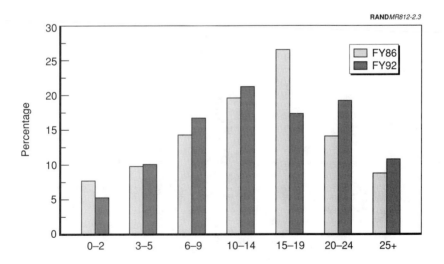

Figure 2.3—Distribution of Officers by Years of Service,
FY86 and FY92

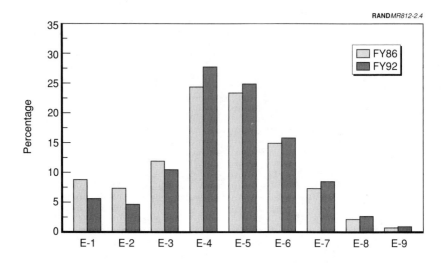

Figure 2.4—Percentage of Enlisted by Paygrade, FY86 and FY92

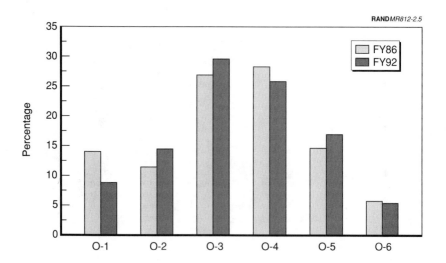

Figure 2.5—Percentage of Officers by Rank, FY86 and FY92

leaving active service, generally with from 4–10 years of service. The trends among officers are a little less clear-cut—we also see a large decline in the proportion of O-1s, a rise in the proportion of O-2s, O-3s, and O-5s (but a puzzling, though small, decline in the proportion of O-4s), and no change among O-6s. These trends at least partly reflect lower levels of officer accessions and perhaps changed promotion policies between ranks.

DEMOGRAPHIC CHARACTERISTICS OF THE SELECTED RESERVE, FY86 AND FY92

Given the data above, it is not surprising to find that the FY92 force is much older than the FY86 force. This is seen in the decline in the proportion of those younger than 25 years and the increase in the proportion of those 40 years and older (Figures 2.6 and 2.7).

There have been some modest shifts in the other demographic characteristics of those serving in the Selected Reserve during this period. The proportion of women in both the officer and enlisted force has increased slightly from 1986 to 1992. However, there is no difference in the proportion of married reservists in the Selected Reserve between 1986 and 1992 (Figure 2.8).

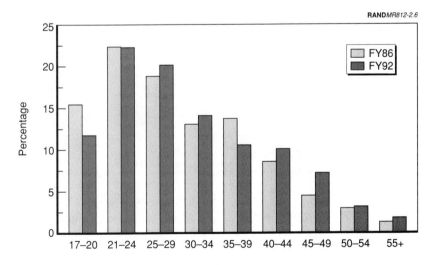

Figure 2.6—Distribution of Enlisted Personnel by Age, FY86 and FY92

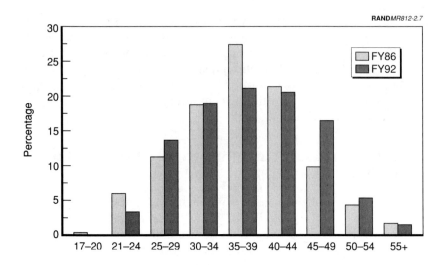

Figure 2.7—Distribution of Officers by Age, FY86 and FY92

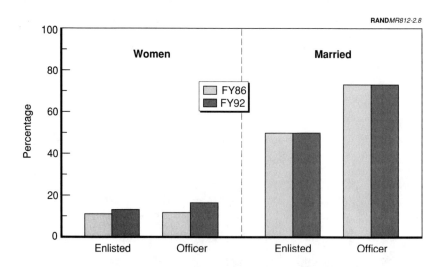

Figure 2.8—Percentage of Women and Married Reservists Among Officer
and Enlisted Personnel, FY86 and FY92

The quality of both enlisted personnel and officers has improved significantly over this time period.

The quality of the enlisted force has been trending upward as shown by the percentage of enlistees having a high school diploma and scoring in the upper third of the Armed Forces Qualifying Test distribution (Category I and II). The percentage of enlisted personnel who are high school graduates has increased by 1–10 percent depending on component, with the USAR and MCR experiencing the largest gains (Figure 2.9). The percentage of enlisted personnel in Category I-II has increased from 38 to 43 percent (Figure 2.10), with the Marine Corps Reserve showing a dramatic rise of almost 14 percent.[2]

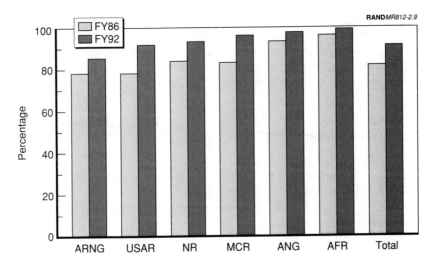

Figure 2.9—Percentage of Selected Reserve Enlisted Personnel Who Are
High School Graduates by Component, FY86 and FY92

---

[2]Cases with missing data were omitted from the calculations. Although all components were missing some level of data on aptitude scores, the ARNG, NR, and MCR had fairly large proportions of cases with missing data. These are reported as a note in Figure 2.13.

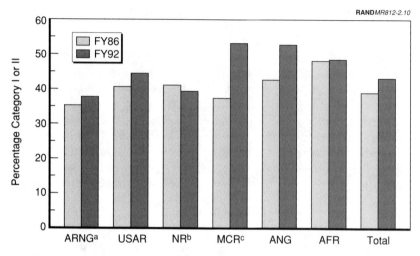

^a10 percent of 1986 sample and 19 percent of 1992 sample were missing data.
^b36 percent of 1986 sample and 33 percent of 1992 sample were missing data.
^c13 percent of 1986 sample and 21 percent of 1992 sample were missing data.

Figure 2.10—Percentage of Selected Reserve Enlisted Personnel in Higher Aptitude Groups by Component, FY86 and FY92

Perhaps the most significant change for officers is the marked increase in the percentage holding college degrees (Figure 2.11). Overall, 75 percent of officers in 1992 were college graduates compared with 67 percent in 1986. The improvement is reflected across each component, with the ARNG, NR, and MCR showing the largest improvements. The low percentage of degree-holders in the National Guard in 1986 reflected mainly their heavy dependence on officer accessions through the state officer programs, which did not require college degrees for entrance. The improvement reflects changing entrance standards for officers into these state programs as well as new policies requiring college degrees to achieve promotions to higher ranks in the reserve.

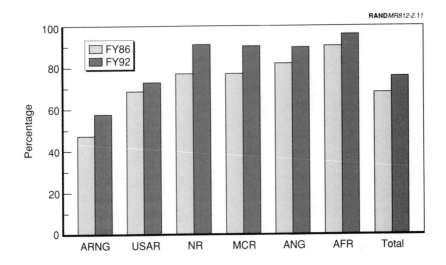

Figure 2.11—Percentage of Selected Reserve Officers with College Degrees
by Component, FY86 and FY92

SUMMARY

Judging by these indicators, the Selected Reserve has become increasingly senior and more experienced from FY86 to FY92. In addition, the quality of the force has also improved significantly and there has been a small increase in the minority representation of the force. These changes need to be kept in mind when we examine differences in the attitudes and perceptions of reservists in the two surveys.

Chapter Three

COSTS AND BENEFITS OF RESERVE PARTICIPATION AND THEIR EFFECT ON RETENTION DECISIONS: 1986 AND 1992

This chapter examines evidence concerning changes in selected monetary and nonmonetary costs and benefits of reserve service and their effect on the decision to stay in the reserve.

An earlier report (Grissmer, Buddin, and Kirby, 1989) outlined in some detail the costs and benefits of reserve participation, pointing out that reservists tend to join and stay in the reserve as long as the long-term monetary and nonmonetary benefits of reserve service exceed the monetary and nonmonetary opportunity costs associated with alternative uses of their time. However, this analysis did not take account of the effects of mobilization. Before ODS/S, few reservists currently in the force had experienced mobilization and thus would not have been able to accurately include a mobilization's effects on their retention decision. If reservists are likely to be called up more frequently or for longer periods of time, this calculus may well change and might potentially cause changing recruiting or retention behavior.

The costs and benefits of reserve service affecting retention decisions are primarily determined by a number of factors related to the following (Grissmer, Kirby and Sze, 1992; Marquis and Kirby, 1989; Grissmer, Buddin and Kirby, 1989; Grissmer, Doering, and Sachar, 1982; Burright, Grissmer, and Doering, 1982):

- Characteristics of the reserve job: motivation for reserve service, reserve pay, years of service, paygrade, component, type of unit, satisfaction with pay, training, and unit;

- Characteristics of the civilian job: employment status, type of employer, pay, whether reservists receive partial/full/no civilian pay during annual training, availability of overtime and opportunity to earn overtime pay, and employer attitude toward reserve service;

- Family support: employment status of spouse, presence of dependents, family problems due to reserve service, and spouse attitude toward reserve service; and

- Characteristics of the individual reservist: age, race/ethnicity, gender, and education.

Here, we show evidence on a more limited subset of variables for two reasons: (a) the primary purpose of this study is to examine the circumstances and environments (family, work, and reserve service) of reservists in 1992 and to the extent possible show how these have changed since 1986; our intention here is *not* estimation of a multivariate reenlistment/retention model because that is the subject of a forthcoming report; and (b) we compare data from the 1992 survey with already published data from the 1986 survey, reported in Grissmer, Buddin, and Kirby (1989). Thus we are limited to the set of previously published variables (although these are quite extensive and include all the major factors).[1]

The comparisons are based on weighted proportions in all cases representing the 1992 and 1986 reserve part-time inventories. Full-time reservists (military technicians and AGRs) were omitted from the analyses because they are full-time civilian employees of the military and their circumstances and perceptions are likely to be quite different from those of part-time reservists.

---

[1]For example, our previous research (Grissmer, Kirby, and Sze, 1992) showed that paygrade, years of service, satisfaction with unit/reserve, presence of dependents, and spouse and employer attitudes were all significantly related to the probability of reenlistment of enlisted reservists making first-term and mid-career reenlistment decisions. We examine all these variables here. As mentioned, we are currently estimating a multivariate model of reenlistment that will take into account the factors listed above.

The sections below examine, in turn, the motivation for reserve service and overall satisfaction with reserve participation, some characteristics of reservists' civilian jobs, and evidence on family support. A final section presents some simple statistics on reenlistment/retention behavior.

Most of the data are disaggregated by paygrade; this allows us to examine differences within each group and also partially controls for the more senior, more experienced force in 1992. However, the sheer volume of the data makes these tables cumbersome to read and digest; as a result, we show only totals in the main text. The complete tables are given in Appendix B. The discussion, however, reflects the changes both in totals as well as within the subgroups.

MOTIVATION FOR RESERVE PARTICIPATION

An indication of the relative importance of factors in the retention decisions of reservists is given in Figures 3.1 and 3.2 (see also Table B.1), which show the percent of enlisted personnel and officers who rated each factor as being of major or moderate importance in the decision to participate/stay in the Guard/Reserve.

The general pattern across the two surveys in terms of relative ranking of the factors important to retention is similar, but there are some potentially significant changes.[2]

- Three reasons rank far above most other reasons for staying and are mentioned by 50 percent or more of reservists as major contributors to staying: retirement benefits, pride in accomplishment, and service to country. In all three cases, the proportions citing these as important are much higher for senior enlisted personnel and this is particularly true in the case of retirement benefits (see Appendix B).

---

[2]In many of the survey comparisons, we do not include tests of whether differences are statistically significant. The large sample sizes of the surveys make almost all comparisons statistically significant, certainly at the aggregate level, and generally at the component level. However, statistical significance does not equate to substantive significance and we focus on those differences that might affect policy. For some comparisons (as, for example, between mobilized and nonmobilized reservist subgroups for which sample sizes are smaller), we do carry out simple tests of differences between population proportions where appropriate.

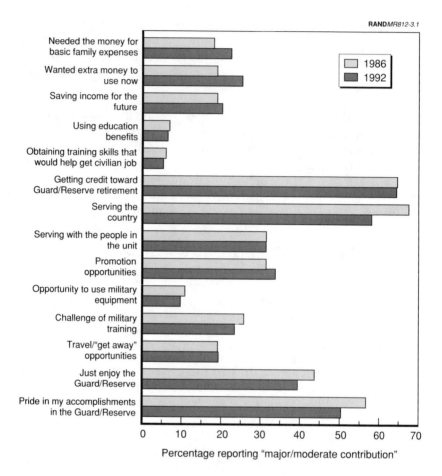

Figure 3.1—Enlisted Personnel: Motivation for Reserve Service,
1986 and 1992

Question: How much have each of the following contributed to your
most recent decision to stay in the Guard/Reserve?

SOURCE: 1992 Reserve Components Survey, Q.30; 1986 Reserve
Components Survey, Q.26.

RAND*MR812-3.2*

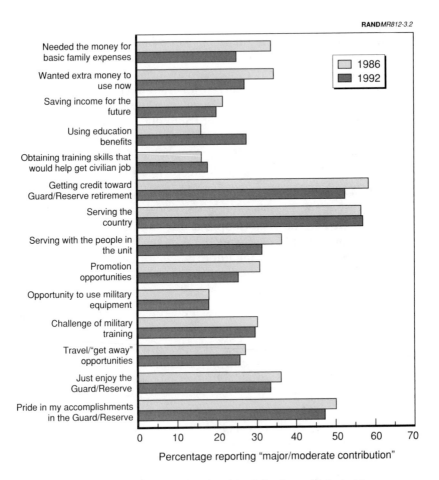

Question: **How much have each of the following contributed to your
most recent decision to stay in the Guard/Reserve?**

SOURCE: 1992 Reserve Components Survey, Q.30; 1986 Reserve
Components Survey, Q.26.

Figure 3.2—Officers: Motivation for Reserve Service,
1986 and 1992

- Interesting changes have occurred in the mention of these three reasons between 1986 and 1992. For officers, pride in accomplishment and serving the country has increased markedly at almost all ranks, whereas retirement as a reason for reenlistment has stayed the same. However, for the enlisted, the pattern is somewhat different. Service to country increased only slightly, whereas pride in accomplishment actually declined. The importance of retirement benefits declined significantly for enlisted personnel. Both of the declines (service to country and retirement benefits) were confined mainly to E-3 and E-4 personnel.

- In 1986, the three next most important reasons cited by nearly one-third of personnel as major contributors to staying in the reserve were serving with people in the unit, "just enjoy the reserve," and current income needs. A large, potentially significant decline has occurred in the importance both officer and enlisted personnel attach to current income needs. This decline appears across nearly all grades.

- Another disparity appears between officer and enlisted in the direction of shift of "just enjoy the reserve" and "serving with people in the unit." For enlisted personnel, each of these declined in importance—mainly for E-3 and E-4 personnel. For officers, these reasons either remained stable or increased in importance.

- A major change between the two surveys is the importance of education benefits in the decision to stay. Formerly, only 16 percent of the enlisted personnel cited this as a major contributor. In the 1992 survey, it increased to 28 percent: an increase of 12 percentage points. For officers, small increases occurred. Not surprisingly, the increases are much larger for lower grades. This is probably a result of the new educational benefits program enacted in 1985, which appears to have attracted college-bound students. This may also help explain the lower importance of current income needs, retirement, importance of serving the country, and serving with people in the unit for junior enlisted. These individuals are oriented to finishing college and probably do not plan on reserve careers or staying with the unit beyond the current term. This underscores the importance of education benefits as a recruiting tool for nonprior-service reservists and its negative effect on retention past the first term.

- Promotions, challenge of military training, and "getting away" were mentioned by between one-fifth and one-third of the reservists. The importance of these seems to have declined slightly over time. The decline is particularly marked among junior enlisted personnel.

- Most of the other factors remained quite stable over the period of the two surveys.

Overall, we can conclude that the motivation for participating and staying in the reserve appears to have shifted somewhat—with less emphasis on immediate compensation and promotion and a greater emphasis on educational benefits. For officers, patriotic and job satisfaction motives increased modestly, whereas for enlisted personnel, these motives were stable or declined slightly. The largest shifts are among junior enlisted personnel who appear to be much more motivated by educational benefits and much less by motivations connected with long-term service in the reserve forces.

GENERAL SATISFACTION WITH RESERVE SERVICE

Additional survey questions deal with overall satisfaction with two aspects of reserve service: military pay and allowances and opportunities for (military) education and training (Figure 3.3; see also Table B.2). Although levels of dissatisfaction are not high, these levels have risen in each category from 1986 to 1992. Among the enlisted, dissatisfaction with military pay and allowances rose from 13 to 18 percent and dissatisfaction with opportunities for (military) education and training rose from 20 to 22 percent. Officers are more satisfied with pay and allowances in both surveys, but even among them we find an increase in dissatisfaction with opportunities for military education and training. In both officer and enlisted ranks, dissatisfaction is much higher among lower-ranking personnel—perhaps due to self-selection.

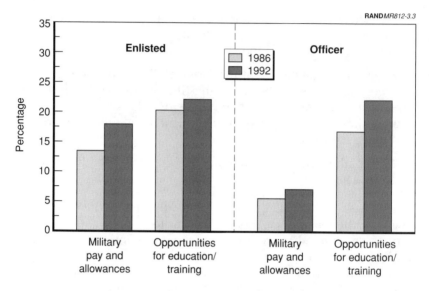

Figure 3.3—Dissatisfaction with Reserve Pay and Training,
1986 and 1992

CHARACTERISTICS OF CIVILIAN JOBS

Some of the key factors determining whether reservists join and reenlist are related to their labor-force participation and the characteristics of their civilian jobs. Reserve service has the potential to conflict with several aspects of civilian activities. First, the time required for reserve service often conflicts with the time demand of their civilian job. Second, reserve service can sometimes reduce the opportunity for advancement and overtime work on the civilian job. Third, reserve service can cause strained relations and conflicts between reservists and their civilian supervisors. We contrast several variables from the 1986 and 1992 surveys to determine if the civilian labor-force activities and civilian job characteristics for reservists have changed in significant ways.

CIVILIAN LABOR-FORCE ACTIVITY AND CIVILIAN EMPLOYER

The labor-force status of reservists in 1986 and 1992 is shown in Table 3.1. The overwhelming majority of reservists are employed either full-time or part-time in the civilian sector. Another 3–6 percent are self-employed. A small proportion of both officers and enlisted are not in the labor force (they are retired, homemakers, or in school) and another small proportion are unemployed and actively looking for work. The labor-force status has changed in two ways: (a) a somewhat larger proportion of both officer and enlisted reservists are not in the labor force; and (b) a small but significant shift of enlisted personnel from full-time to part-time work can be seen. These shifts are partially related to two factors: the larger numbers of college students in the enlisted ranks and the somewhat higher number of women in the officer ranks. Both of these groups are more likely to have part-time jobs or to be out of the labor force. Also, the reserve forces initiated a college tuition program in 1985, which provides significant payments for college tuition and expenses, and this program has probably attracted more college students into the enlisted ranks. This increased participation from college students may also help explain the higher aptitude levels of junior enlisted.

Table 3.1

Employment Status of Part-Time Reserve Members, 1986 and 1992

| | Percent of Total | | | |
| | Enlisted | | Officers | |
Civilian Employment Status	1986 Survey	1992 Survey	1986 Survey	1992 Survey
Working full-time	73	69	80	79
Working part-time	10	12	6	5
Self-employed	3	3	7	6
Unemployed	7	6	2	2
Not in labor force	6	11	5	7

SOURCE: 1992 Reserve Components Survey, Q.106B–106L; 1986 Reserve Components Survey, Q.3, Q.93M.

Table 3.2 shows the distribution of all employed reservists by type of employer for both 1986 and 1992. Again, the distributions are remarkably similar. Thirty percent of part-time enlisted reservists and 40 percent of officers work in the public sector. There has been a slight increase in the number of officers working for the federal government. Perhaps more significantly, there has been a decline in the proportion of self-employed enlisted personnel from 1986 to 1992. If self-employed reservists were more likely to lose income during mobilization, then they might be expected to have lower retention in the wake of ODS/S.

Table 3.2

Employer by Type and Size: Part-Time Reserve Members, 1986 and 1992

	Percent of Total			
	Enlisted		Officers	
Type of Employer	1986 Survey	1992 Survey	1986 Survey	1992 Survey
Public sector				
Federal government	10	10	14	17
State government	8	9	11	10
Local government	9	9	10	10
Private sector				
Firm < 500 employees	35	37	20	20
Firm ≥ 500 employees	27	29	33	34
Self-employed/family business	11	6	11	10

SOURCE: 1992 Reserve Components Survey, Q.110; 1986 Reserve Components Survey, Q.3, Q.97.

Employer Pay Policies and Overtime Opportunity

One important cost of reserve participation is forgone civilian income during military leave (generally because of needing to attend annual training and drills). Reservists will lose income during annual training if their combined civilian income and military pay for that period does not meet their full civilian income. Each employer has a policy mandating full, partial, or no civilian pay for reserve duty. The partial pay policy is usually set to make up any differences between civilian and military income. This policy allows reservists to essentially break even during annual training.

Earlier research (Grissmer, Buddin, and Kirby, 1989) shows that junior personnel lose almost as much as senior personnel in absolute terms despite the fact that senior personnel tend to have significantly higher income. This is balanced by the much higher probability of senior personnel receiving partial or full wages from civilian employers.

Figures 3.4 and 3.5 (Table B.3) show employer pay policies for reservists on military leave for annual training in 1986 and 1992. A significant shift has occurred for enlisted personnel toward less civilian compensation. In 1986, about 47 percent of employers provided no civilian income compared to 52 percent in 1992. For officers, a shift toward partial pay has occurred with small declines in both the full pay and no pay categories. For both officer and enlisted, reserve service occurred more often on work days in 1992 than in 1986. The latter may be a result of higher employment, a healthier economy, or increased military demands on reservists in 1992.

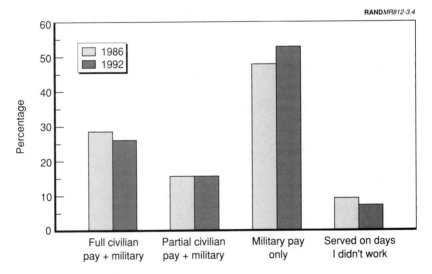

Figure 3.4—Civilian Employers' Pay Policies for Annual Training,
1986 and 1992: Enlisted Personnel

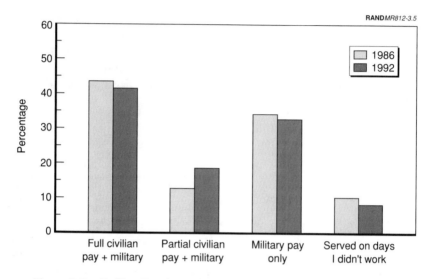

Figure 3.5—Civilian Employers' Pay Policies for Annual Training,
1986 and 1992: Officers

Apart from forgone income, another monetary opportunity cost may arise because the reserve job may involve forgoing the opportunity to work overtime hours on the civilian job or to work at another moonlighting job. Occasionally, the reservist may be passed up for civilian promotion. These opportunity costs vary a great deal among reservists.

Figures 3.6–3.9 (Tables B.4–B.5) examine the issue of overtime opportunity and wage premiums for overtime. Both the 1986 and 1992 survey data indicate that a significant number of enlisted personnel (approximately 45 percent) lose some overtime opportunity and wages as a result of the reserve job. This is not as much of an issue for officers, presumably because many of them (especially the more senior officers) work in professional jobs that are classified "exempt" and thus are not eligible for overtime. Lost overtime is much more prevalent among lower-ranking enlisted and officer personnel. Fewer senior personnel are also more likely to be paid wage premiums for overtime: For example, 60–65 percent of younger enlisted personnel and 40 percent of young officers receive premiums for working overtime.

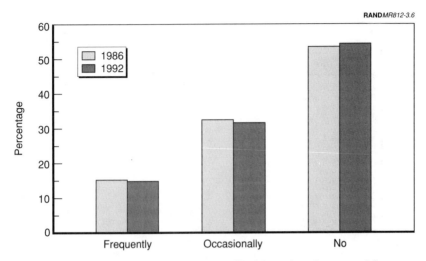

Figure 3.6—Percentage Losing Civilian Overtime Opportunities,
1986 and 1992: Enlisted Personnel

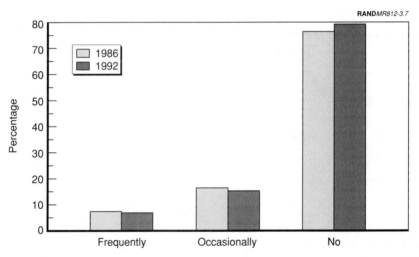

Figure 3.7—Percentage Losing Civilian Overtime Opportunities,
1986 and 1992: Officers

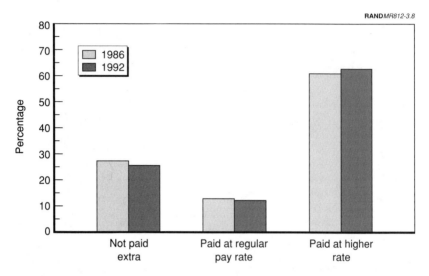

Figure 3.8—Civilian Overtime Pay Rate, 1986 and 1992: Enlisted Personnel

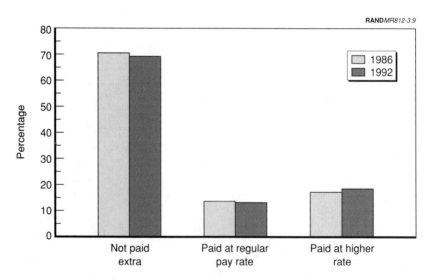

Figure 3.9—Civilian Overtime Pay Rate, 1986 and 1992: Officers

There is little change from 1986 to 1992 in lost overtime opportunity, but there is a slight shift upward in the proportion of officers and enlisted receiving premium overtime pay rates. So, should more overtime opportunities arise, reservists may have slightly higher opportunity costs.

Attitudes of Civilian Supervisors Toward Reserve Service

Earlier research (Grissmer, Kirby, and Sze, 1992) has shown that the most important civilian job characteristic associated with retention is the negative attitude of the civilian supervisor. Conflicts with supervisors over reserve duty can potentially result in high opportunity costs in terms of promotions and job satisfaction. Figures 3.10 and 3.11 (Table B.6) provide evidence on supervisors' perceived attitudes toward reserve service. The data from the two surveys indicate a significant reduction in very unfavorable attitudes among supervisors with corresponding increases in very favorable attitudes. These changes appear at nearly all grades.

One explanation for this trend is a revised perception of the "value" of reserve service among the civilian population. Before ODS/S, the reserve forces were probably seen as a more marginal component of defense forces, whereas after ODS/S, reserve forces are more readily accepted as integral to national defense.

Figures 3.12 and 3.13 (Table B.7) examine the incidence of employer conflicts arising from attendance at reserve drills and annual training or from extra time spent on Guard/Reserve business. Attendance at annual training appears to cause the most problems, with about 30 percent of both officers and enlisted reporting that such absence was a serious/somewhat serious problem for their employers, followed by extra time spent at Guard/Reserve (20–30 percent, depending on rank and survey). Officers in general report a somewhat higher incidence of serious employer problems as a result of reserve participation (with the exception of weekend drills).

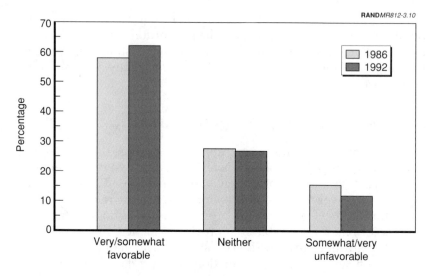

Figure 3.10—Attitude of Civilian Supervisor Toward Guard/Reserve
Participation, 1986 and 1992: Enlisted Personnel

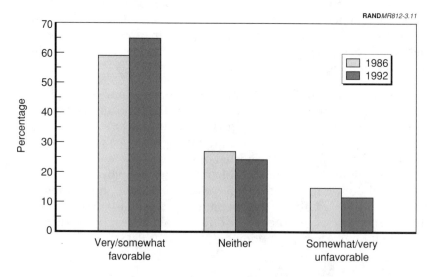

Figure 3.11—Attitude of Civilian Supervisor Toward Guard/Reserve
Participation, 1986 and 1992: Officers

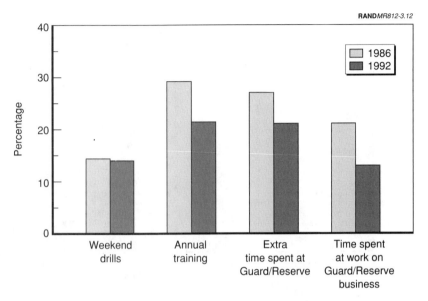

Figure 3.12—Employer-Related Problems Resulting from Reserve Service,
1986 and 1992, Enlisted Personnel: Percentage Reporting a
"Serious/Somewhat Serious" Problem

Overall, these data show significant shifts toward fewer conflicts between reserve service and civilian jobs. In addition, employers appear to be more tolerant of extra time spent at the Guard/Reserve and time spent at work on Guard/Reserve business (assuming that the frequency of this has not changed over time) as witnessed by the smaller numbers of reservists reporting that these activities caused serious problems for employers.

CONFLICTS WITH FAMILY TIME AND SPOUSE ATTITUDES

For many reservists, an important component of the cost of reserve participation is the decrease in the amount of time available to spend with their families or in leisure pursuits.   This, of course, may have an effect on spouse attitudes that previous research (Grissmer, Kirby, and Sze, 1992) has shown to be a key variable in reenlistment behavior. Responses to questions regarding reservists' use of time are shown in Table B.8. By and large, reservists feel that they spend sufficient time on both civilian jobs and reserve jobs but that time spent

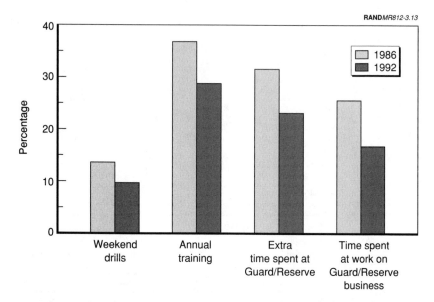

RAND*MR812-3.13*

**Figure 3.13—Employer-Related Problems Resulting from Reserve Service,
1986 and 1992, Officers: Percentage Reporting a "Serious/Somewhat
Serious" Problem**

on family, leisure, and community activities is insufficient. Between
60 and 75 percent feel that time spent on family and leisure activities
is insufficient, whereas 50–58 percent feel that they do not spend
enough time on community activities compared to the 5–10 percent
who feel they do not spend enough time at their civilian or reserve
jobs. Figure 3.14 shows the percentage of reservists who feel that
they spend too much time on the reserve job, which ranges from 8–
20 percent. The good news is that this number has declined from
1986 to 1992; this is at least partly the result of the increased full-time
support provided by the components in recognition of the pressures
faced by their officers and senior enlisted in planning training and
attending to administrative details.

Officers and senior enlisted personnel appear to be most sensitive to
time pressures that prevent family or leisure activities. Obviously,
this pattern of dissatisfaction is not unique to reserve service, although inflexible schedules and other demands of reserve service

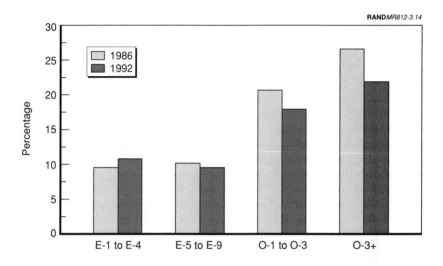

Figure 3.14—Percentage of Reservists Who Feel That They "Spend
Too Much Time on Reserve Job," 1986 and 1992

may well contribute to or exacerbate such feelings. Junior enlisted
personnel report higher levels of "not spending enough time on family and leisure time" in 1992 compared to 1986. There is little significant change between 1986 and 1992 in the remaining variables.

The conflict between family and reserve time is shown in Figures 3.15
and 3.16 (Table B.9. For enlisted personnel, the data show about the
same level of problem for weekend drills, slightly increased values for
annual training, and lower levels of problems for extra time. About a
quarter of the enlisted encounter problems with time for annual
training and 16 percent of enlisted report problems with weekend
drills. In general, younger enlisted personnel and officers tend to
face more problems than older, more experienced personnel.

Among officers, the percentage reporting serious/somewhat serious
family problems because of reserve service has declined over time,
significantly so in the case of extra time spent at Guard/Reserve
training. In 1992, 28 percent of officers reported family problems
because of extra time spent on Guard/Reserve duties; this is
appreciably lower than the 39 percent reported in the 1986 survey.
Two reasons can be suggested for the change: Perhaps the reserve

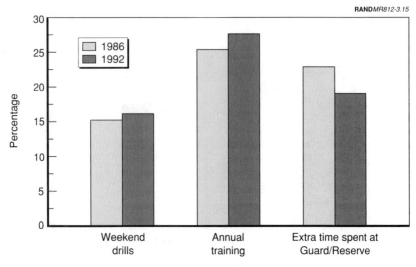

Figure 3.15—Family Problems Resulting from Reserve Service, 1986 and
1992, Enlisted Personnel: Percentage Reporting a "Serious/Somewhat
Serious" Problem

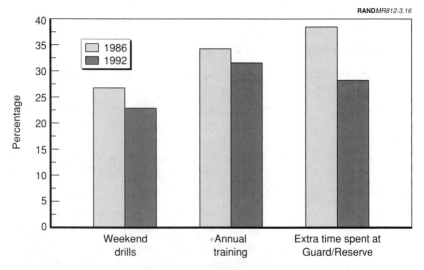

Figure 3.16—Family problems Resulting from Reserve Service, 1986 and
1992, Officers: Percentage Reporting a "Serious/Somewhat
Serious" Problem

has become more sensitive to this issue given the 1986 data and has made attempts to lessen this burden and/or (as suggested above) the increase in full-time support has reduced the administrative and other workload faced by part-time reservists, especially officers.

Reservists were also asked about the overall attitude of their spouses toward participation in the Guard/Reserve. Table B.12 lists these responses by paygrade; overall percentages are shown in Figures 3.17 and 3.18. Although most spouses have generally favorable attitudes toward reserve service, an unfavorable spouse attitude affects 12 percent of the enlisted and 11 percent of officers. This proportion has remained essentially constant over the six years between the two surveys. As can be expected, younger officers and enlisted face a higher incidence of unfavorable attitudes. What makes these data important is the evidence presented in earlier RAND work (Grissmer, Kirby, and Sze, 1992) that attitudinal variables appeared to exert a strong influence on the decision to reenlist among enlisted personnel. Perceived spouse attitude, for example, was the most important predictor of reenlistment probability for reservists in the

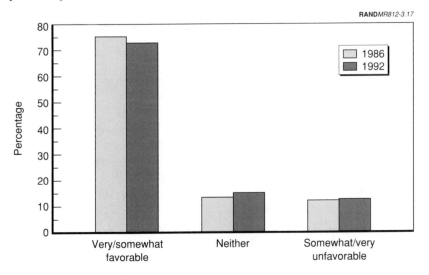

Figure 3.17—Spouse Attitudes Toward Reserve Participation, 1986 and 1992: Enlisted Personnel

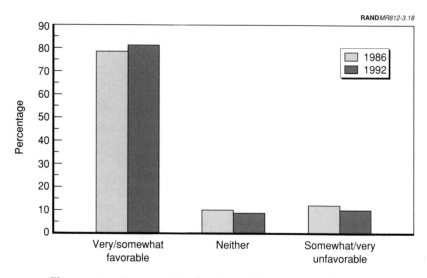

Figure 3.18—Spouse Attitudes Toward Reserve Participation,
1986 and 1992: Officers

1986 survey.[3] Spouse attitudes toward reserve service seem to have
been largely unchanged by ODS/S.

FUTURE REENLISTMENT PLANS/RETENTION BEHAVIOR

Decisions to stay in the reserve depend on the relative weights individuals place on the benefits and opportunity costs of participation.
Reservists were asked how likely they were to reenlist or extend at the
end of their term of service (ETS). Table 3.3 reports the percentage of
those reporting "No Chance" and "Slight/Very Slight Possibility" of
reenlistment/continuation. We have comparison numbers for 1986
for the enlisted only but report the 1992 survey results for both officer
and enlisted personnel. Overall there appears to be a significant

---

[3]Clearly, we need to caveat these results to some extent. The reported attitude of the
spouse may be a reflection of the individual's own satisfaction/dissatisfaction. However, there was some evidence from the question in the survey regarding the individual's own overall level of satisfaction/dissatisfaction with his Guard/Reserve participation that the perceived attitude of the spouse was indeed an independent variable
or at least was measuring a dimension other than the reservist's own feelings.

Table 3.3

Reenlistment/Continuation Intention of Reservists

Question: How likely are you to reenlist or extend at the end of your current term of service? (enlisted). At the completion of your obligation or term of service, how likely are you to continue to participate in the Selected Reserve of the Guard/Reserve? (officers). (Percentages shown are for those reporting "no chance" and "very slight/slight possibility")

	1986 Survey		1992 Survey	
Grade	No Chance	Very Slight/ Slight Possibility	No Chance	Very Slight/ Slight Possibility
Enlisted				
E-3	19.2	18.0	27.1	16.3
E-4	10.5	13.7	16.7	14.9
E-5	5.1	7.9	10.2	8.7
E-6	5.2	6.6	11.7	8.7
E-7	7.1	6.7	17.2	10.8
E-8	8.4	5.0	18.2	12.3
E-9	20.1	6.3	26.3	7.3
Total	7.6	9.4	15.3	11.8
Officer				
O-1	(a)	—	4.5	10.7
O-2	—	—	8.1	11.4
O-3	—	—	6.3	12.8
O-4	—	—	5.8	6.6
O-5	—	—	20.6	9.6
O-6	—	—	19.1	10.1
Total	—	—	7.6	11.3

SOURCE: 1992 Reserve Components Survey, Q.23 and Q.22; 1986 Reserve Components Survey, Q.18, 19, 20.

aData not available.

increase in those who report that their chance of reenlisting/cointinuing is small. In 1986, 17 percent reported a small or no chance of reenlistment/continuation compared to 27 percent in 1992. This increase occurs across all paygrades. The group reporting no chance of reenlisting/continuing has doubled from 1986 to 1992, and there is also a small increase in the proportion reporting that there is only a slight possibility of reenlisting/continuing. Some of the increase reported in higher paygrades can be explained by the fact that many more of them are probably retirement-eligible, given the increase in years of service. However, this does not account for the large increase among the junior paygrades.

Officers in general appear much more likely to stay than the enlisted, with only 8 percent reporting "no chance" of continuing. The percentage increases to 20 percent among the higher ranks presumably because they are more likely to be eligible for retirement.

Although intentions to reenlist/continue have been shown to be correlated with actual reenlistment decisions, earlier data from the 1986 survey showed that almost all individuals underestimate their probability of staying, with the exception of those who were almost certain to stay (Grissmer, Kirby, and Sze, 1992). Among the enlisted, we found that the discrepancy between the subjective probability and actual behavior was quite large (particularly for those with low probabilities).[4]   For instance, those who stated their probability of reenlistment to be 0.10 had an actual reenlistment rate of 0.57.

Another report will examine in more detail retention behavior in the wake of ODS/S.   In this report, we compare stated reenlistment/ continuation intentions with whether the individual was there at the end of FY94 to provide some indication as to whether intention still is correlated with behavior. Figure 3.19 shows results similar to those in the earlier research—a positive correlation between intention and behavior—but it shows that reservists underestimate their actual retention behavior.

There are fairly large differences in the retention rate of individuals who claim that there is little or no chance of staying and reservists in general. From 20–25 percent of the reservists left between the time the survey was fielded and the end of FY94. However, among those with low reenlistment/continuation probabilities, the percentage remaining was considerably smaller. Only about 45 percent of enlisted personnel and 57 percent of these officers remained two years after the survey.

It is also interesting that not all of them have left, despite being certain that they would not stay. We can offer a number of reasons for this:

------

[4]Enlisted reservists in the 1986 survey were tracked forward through September 1987 (which gave us an elapsed time of 6–18 months from the completion of the survey) to see whether they were still in the Selected Reserve. We then compared the actual reenlistment rate for each group.

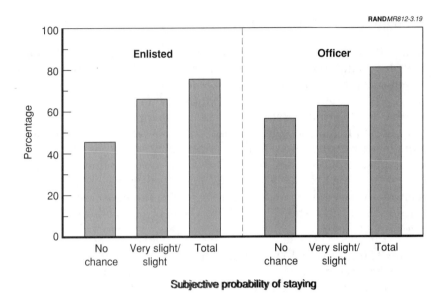

RAND*MR812-3.19*

Figure 3.19—Stated Reenlistment /Continuation Intentions in 1992
Compared with Percentage Remaining in the Selected Reserve
in September 1994

- Given that a considerable period of time has elapsed between the time of the survey (one to two years) and the behavior being examined, it is likely that some of the factors negatively influencing the intention to reenlist/continue may have changed for the better (employer or spouse conflicts, problems with the unit, etc.)

- The reserves may offer some very effective intervention or counseling at the time of the decision, so that individuals who are disgruntled are persuaded to stay.

- Some individuals may not have reached reenlistment/ continuation points yet, so the retention rate does not reflect the reenlistment/continuation decision.

- The data may reflect a "protest effect," where individuals with serious complaints against the system may underestimate their likelihood of staying; however, at the time of the actual decision, other factors may outweigh the negative ones.

1986 AND 1992 RETENTION DATA

In an effort to see whether changing circumstances and motivation have led to different patterns of retention, we compared published retention data for 1986 and 1992.[5] Some care is required in comparing retention data because of the "stop-loss" policies implemented during ODS/S. After reserve mobilization, a stop-loss policy was instituted to prevent personnel whose term of enlistment had ended from leaving the Selected Reserve. This policy would have the effect of raising retention rates during ODS/S but lowering them after the end of ODS/S when those whose ETS decisions were deferred were allowed to leave. ODS/S ended in March 1991 for most reservists. As a result, FY92 data should be relatively uncontaminated by ODS/S personnel policies.[6]

Figure 3.20 compares FY86 continuation data with FY92 data for Selected Reserve officers and enlisted personnel in all components combined. We divide enlisted personnel and officers into groups with less than six years of service and more than six years; among enlisted personnel, these two groups are generally referred to as "first-termers" and "careerists." Officer and enlisted continuation rates have remained remarkably stable over the two time periods, with a very slight dip for the more senior group.

Figures 3.21–3.24 show continuation rates for enlisted personnel and officers for each of the components separately. All components have higher enlisted first-term continuation rates in 1992 than in 1986; among officers, there is no change with the exception of the NR, which shows a drop of 5 percent. Career continuation rates have declined slightly for enlisted personnel in all components, as have the continuation rates among officers with more than six years of service.

---

[5]Data are from the Official Guard and Reserve Manpower Strengths and Statistics, RCS: DD-RA(M)1147/1148, Assistant Secretary of Defense (Reserve Affairs) for FY86 and FY92.

[6]Buddin and Kirby (1996) analyzed in more detail retention and attrition data from 1986–1994. Their basic finding was that ODS/S did not dramatically change attrition and retention behavior.

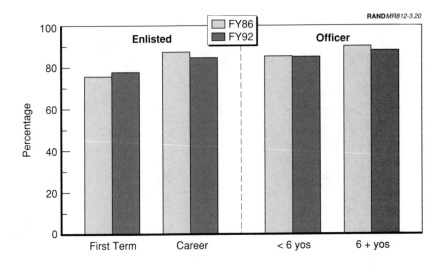

Figure 3.20—Officer and Enlisted Continuation Rates, FY86 and FY92

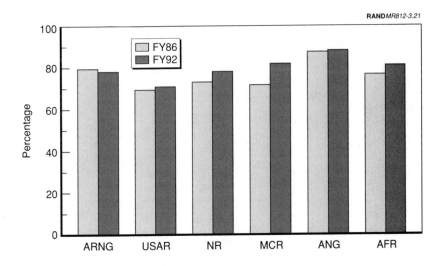

Figure 3.21—Enlisted First-Term Continuation Rates by Component,
FY86 and FY92

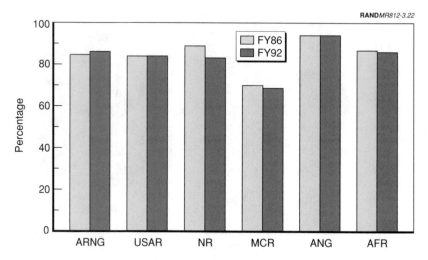

Figure 3.22—Continuation Rates of Officers with Less Than Six Years
of Service by Component, FY86 and FY92

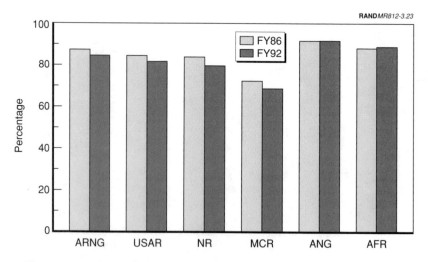

Figure 3.23—Continuation Rates of Enlisted Careerists by Component,
FY86 and FY92

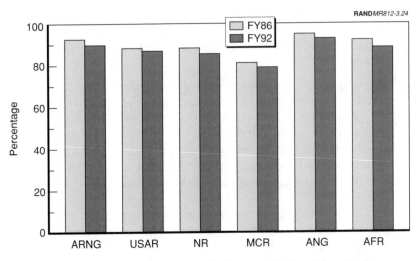

Figure 3.24—Continuation Rates of Officers with More Than Six Years
of Service by Component, FY86 and FY92

SUMMARY

Overall, the attitudes and characteristics of reservists and their work and family environments show great stability from 1986 to 1992, with respect to many of the variables that could be expected to affect retention. Indeed, most of the changes, while not dramatic, would tend to have positive effects on retention.

- The motivation for staying in the Guard/Reserve appears to have shifted from 1986 to 1992: Among enlisted personnel, there is less emphasis on immediate compensation and promotion and greater importance placed on educational benefits; among officers, patriotic and job satisfaction motives were more frequently mentioned. Expanded educational benefits may have attracted a newer group of young enlisted personnel whose primary motivation is using educational benefits while attending college, not long-term reserve careers.

- Offsetting the decline in importance of immediate compensation as a factor in retention, we see a small but definite increase in levels of dissatisfaction with military pay and opportunities for education/training among both officers and enlisted. Part of the dissatisfaction with pay may be a reflection of the perceived higher risk of mobilization and the attendant likely economic losses.

• Perhaps the most important positive change is the shift in employer attitudes from 1986 to 1992. There is a small increase in the proportion of reservists reporting a favorable attitude on the part of their civilian supervisors. In addition, reservists report much less conflict with employers in attending drills and annual training and in spending extra time on reserve obligations while on the job. (In the latter case, this may be due to less demand from the reserve job for extra time.) ODS/S highlighted the significant contributions reservists made to the war effort and underscored their importance to the total force. This probably caused employers to become more favorably disposed toward their reserve participation and lessened the inevitable conflicts that arise between the demands of the reserve and the civilian jobs.

• Offsetting this to some extent is the small change in employer pay policies during reserve service. There is a small decline in the proportion of enlisted who received full civilian pay during annual training and an increase in those reporting that they received only military pay. Among officers, the news is mixed. Employers appear to have moved away from offering either full or no civilian pay in favor of offering partial compensation.

• Perceived attitudes of spouses have remained stable between 1986 and 1992 (a little surprising given the increased chance of mobilization). These data are particularly important because of the importance of spouse's attitude in reenlistment decisions.

• Reservists across most grades reported much lower subjective probabilities of reenlistment/continuation in the 1992 survey than in the 1986 survey. However, preliminary evidence from simple continuation rates shows that rates for those with less than six years of service have changed little between 1986 and 1992 for both officer and enlisted personnel, but continuation rates among those with more than six years of service have declined slightly. The data show no dramatic change in overall behavior that could be attributable to ODS/S. However, we must caution that this finding is based on aggregated, very simple measures. Much further work remains to be done before we can reach any such conclusion definitively.

Chapter Four

DIFFERENCES IN SPOUSE AND EMPLOYER ATTITUDES AND SATISFACTION WITH RESERVE PARTICIPATION BETWEEN MOBILIZED AND NONMOBILIZED RESERVISTS

Manpower planners are concerned with the key question of whether participation in ODS/S affected those who were mobilized in significant ways. For example, did they face greater family and employer conflicts because of being called up? Do mobilized reservists have different attitudes toward their reserve job that could affect retention and morale? Given the increased frequency with which reservists are being called up for peacetime missions, answers to these and related questions are crucial. We use the 1992 survey data to shed some light on these questions by comparing mobilized and nonmobilized reservists.

Appendix D compares the demographic and service-related characteristics of mobilized and nonmobilized reservists. Because the mobilized group is composed of a somewhat higher grade distribution than the nonmobilized group, we disaggregate all the data shown here by rank.

DIFFERENCES IN SPOUSE AND EMPLOYER ATTITUDES

Figures 4.1–4.4 depict the differences in spouse and employer attitudes for mobilized and nonmobilized enlisted personnel and officers. For each grade, we show the percentage reporting an unfavorable attitude toward reserve participation on the part of the spouse

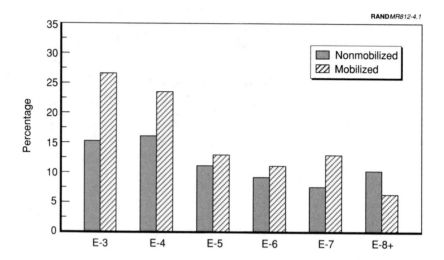

Figure 4.1—Percentage with Spouse with an Unfavorable Attitude
Toward Guard/Reserve Participation: Mobilized and
Nonmobilized Enlisted Personnel, 1992

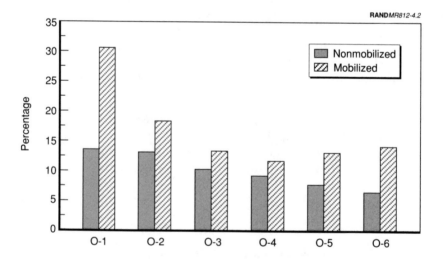

Figure 4.2—Percentage with Spouse with an Unfavorable Attitude
Toward Guard/Reserve Participation: Mobilized and
Nonmobilized Officers, 1992

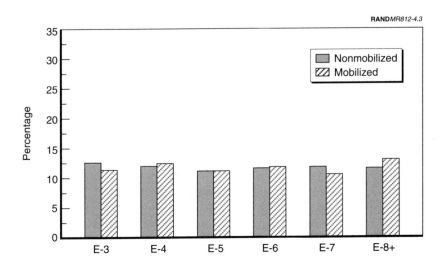

Figure 4.3—Percentage with Civilian Supervisor with an Unfavorable
Attitude Toward Guard/Reserve Participation: Mobilized and
Nonmobilized Enlisted Personnel, 1992

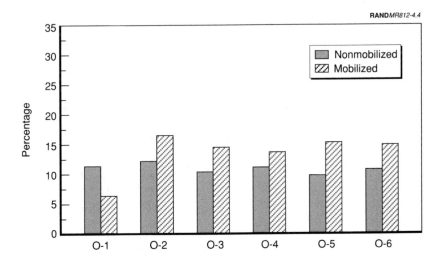

Figure 4.4—Percentage with Civilian Supervisor with an Unfavorable
Attitude Toward Guard/Reserve Participation: Mobilized and
Nonmobilized Officers, 1992

or the employer. The graphs are on the same scale to facilitate comparison and to show the relative magnitude of the two problems.

The percentage of reservists reporting that their spouses have unfavorable attitudes toward their Guard/Reserve participation is significantly higher for the mobilized reservists than for the nonmobilized reservists—especially in the case of the more junior grades. This is true for both enlisted personnel and officers. For example, whereas only about 15 percent of the nonmobilized E-3s and E-4s report that their spouses have an unfavorable attitude, over a quarter of the mobilized junior grades do so. Over 30 percent of mobilized O-1s fall into this category compared to less than 15 percent of the nonmobilized group. If this is being triggered by mobilization, it may well have effects on subsequent retention and possibly recruiting.

On a more positive note, there is little or no difference in civilian supervisors' attitudes toward mobilized and nonmobilized reservists among the enlisted personnel. However, all officer ranks report an increase in the incidence of unfavorable employer attitudes, with the exception of O-1s, who show a marked decline. Given the data in the last chapter showing overall positive shifts between 1986 and 1992, it appears that the positive shift was primarily for employers of nonmobilized officers.

SATISFACTION WITH THE GUARD/RESERVE AND FUTURE INTENTIONS

We now examine satisfaction/dissatisfaction with pay and benefits and overall participation in the Guard/Reserve. Figures 4.5–4.8 present the percentages of enlisted personnel and officers who are very/somewhat dissatisfied with these aspects of the Guard/Reserve. Overall, only 10–20 percent of enlisted personnel and 5–15 percent of officers reported being very or somewhat dissatisfied with pay/ benefits in the Guard/Reserve. Among both officers and enlisted personnel, junior members tended to be more dissatisfied with pay and benefits than more senior grades, and junior members who were mobilized had the highest levels of dissatisfaction. For example, one-fifth of E-4s who were mobilized reported being very dissatisfied with reserve pay and benefits compared to only 15 percent of the nonmobilized E-4s. The difference is marked in the case of O-1s and

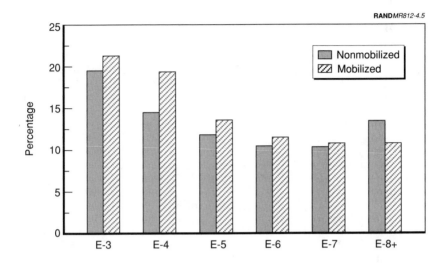

Figure 4.5—Percentage Dissatisfied with Reserve Pay and Benefits:
Mobilized and Nonmobilized Enlisted Personnel, 1992

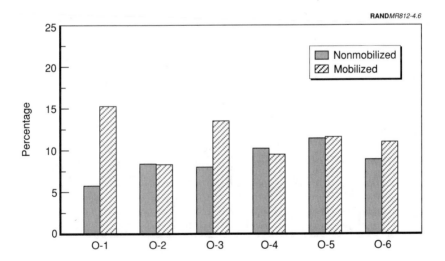

Figure 4.6—Percentage Dissatisfied with Reserve Pay and Benefits:
Mobilized and Nonmobilized Officers, 1992

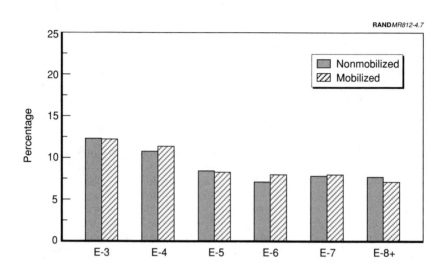

Figure 4.7—Percentage Dissatisfied with Overall Participation in the
Guard/Reserve: Mobilized and Nonmobilized Enlisted Personnel, 1992

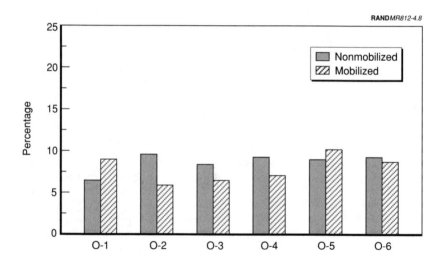

Figure 4.8—Percentage Dissatisfied with Overall Participation in the
Guard/Reserve: Mobilized and Nonmobilized Officers, 1992

O-3s (but not O-2s). Senior officers and enlisted tend to show little difference in levels of satisfaction between mobilized and nonmobilized personnel (notice the reversal in the case of the E-9s).

It is surprising that when we consider overall dissatisfaction with reserve service, we find little difference at any rank between mobilized and nonmobilized enlisted reservists (Figure 4.7). About 10 percent or less express dissatisfaction. For officers, dissatisfaction levels are even lower, with O-2 to O-4 mobilized officers expressing *less* dissatisfaction then their nonmobilized counterparts.

To see whether high levels of dissatisfaction and/or unhappiness over the ODS/S experience translated into a determination to leave the reserve, we examined the intention to reenlist among the two groups. Figures 4.9 and 4.10 show the percentage of reservists who were certain that they were not going to reenlist in the Guard/ Reserve once their current term of service was completed. Among junior enlisted grades, we find that higher proportions of E-3s and E-4s among the mobilized group were certain that they would leave the Guard/Reserve than in the nonmobilized group. The trend is different among officers, where nonmobilized reservists reported a much higher probability of not reenlisting.

Tables 4.1 and 4.2 break out some of the reasons why individuals may want to leave the Guard/Reserve. Ineligibility to reenlist, slowness of promotions, conflicts between attendance at unit drills and civilian job, desire for more leisure time or time to spend with family all head the list, whereas problems caused by mobilization appear to be much lower on the list.

As in the previous chapter, we were curious to see how well intentions tracked behavior. We looked to see whether the individual who stated that he was certain he would not reenlist or continue was still in the reserves approximately two years after the survey. Figure 4.11 shows the proportion of enlisted personnel and officers with a zero or very slight probability of reenlisting/continuing who were still in the Reserve Components at the end of FY94. (Depending on when reservists were given the survey, the latter date is between 24 and 30 months after the survey.)

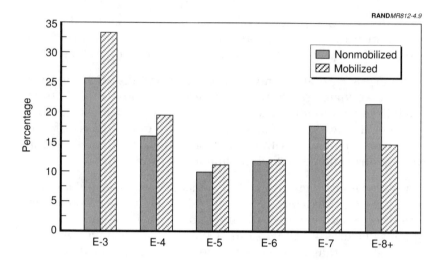

Figure 4.9—Percentage Certain Not to Reenlist: Mobilized and
Nonmobilized Enlisted Personnel, 1992

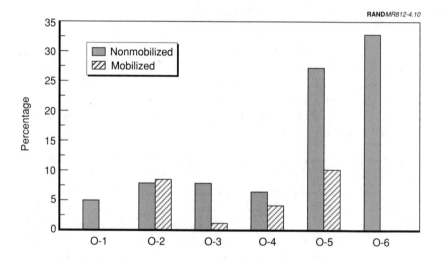

Figure 4.10—Percentage Certain Not to Continue: Mobilized and
Nonmobilized Officers, 1992

Table 4.1

Reasons for Not Reenlisting in the Guard/Reserve: Enlisted Personnel

Reason	Nonmobilized		Mobilized	
	Most Important Reason (%)	Second Reason (%)	Most Important Reason (%)	Second Reason (%)
I am not eligible to reenlist	17.4	4.9	17.1	4.3
I am moving to another area	6.4	7.6	5.8	7.5
It is too hard to get to my Guard/Reserve unit	2.6	3.9	2.5	3.3
I need the time for my education	5.6	4.6	6.1	4.6
My unit drills conflict with my civilian job	14.1	11.8	13.4	12.5
My unit drills conflict with my family activities	11.8	12.7	10.0	10.9
I want more leisure time	5.6	10.2	5.0	8.2
I don't like my unit's training	4.5	5.7	4.3	6.0
My unit doesn't have modern equipment for training	3.6	5.4	2.9	4.3
I'm bored with unit activities	5.6	8.6	4.8	9.2
The pay is too low	3.4	6.5	3.2	6.0
Promotions are too slow	16.8	14.0	19.0	14.1
I've had too many problems getting paid	1.2	2.0	1.4	3.2
Problems caused by mobilization/activation/deployment	1.5	2.2	4.4	6.0

SOURCE: 1992 Reserve Components Survey, Q.24A–N, p. 6.

There is little difference between the mobilized and nonmobilized enlisted personnel in terms of retention. The overall retention rate is 75 percent; among those certain not to reenlist, it is significantly lower, close to 55 percent. Some of this is being driven by retirements of older personnel; however, among the junior grades, we find a difference in the two-year retention rate of 4–7 percent, with mobilized personnel leaving at higher rates. Among officers who were certain that they would not stay in the Guard/Reserve, we find a sizable difference between the retention rate of mobilized and nonmobilized officers, with mobilized officers leaving the reserves at much higher rates than nonmobilized officers. Again, some of this is being driven by retirements; however, among O-1s, there is a

Table 4.2

Reasons for Not Continuing in the Guard/Reserve: Officers

	Nonmobilized		Mobilized	
	Most Important Reason	Second Reason	Most Important Reason	Second Reason
Reason	(%)	(%)	(%)	(%)
I am not eligible to reenlist[a]	14.5	3.2	14.0	2.1
I am moving to another area	7.9	7.6	9.7	8.0
It is too hard to get to my Guard/Reserve unit	4.5	6.8	4.6	6.6
I need the time for my education	3.0	3.7	2.9	3.4
My unit drills conflict with my civilian job	19.3	17.5	15.9	15.6
My unit drills conflict with my family activities	22.2	20.0	20.4	19.7
I want more leisure time	9.4	17.0	8.2	14.2
I don't like my unit's training	3.1	3.3	2.6	3.2
My unit doesn't have modern equipment for training	1.8	2.5	1.5	2.5
I'm bored with unit activities	4.2	5.4	3.1	4.7
The pay is too low	1.5	2.6	1.9	3.0
Promotions are too slow	5.3	5.9	6.4	6.7
I've had too many problems getting paid	1.0	1.5	1.6	3.0
Problems caused by mobilization/activation/deployment	2.4	3.1	7.3	7.4

SOURCE: 1992 Reserve Components Survey, Q.24A–N, p. 6.

[a]We should point out that because officers to not "reenlist," some may have misunderstood the question. As a result, these data need to be interpreted with caution

difference of 7 percent in retention rates of mobilized and nonmobilized personnel. This trend, if borne out by further analysis, bears observation.

We must emphasize the limitations of the current analysis. We have not controlled for ETS decision points, nor are we able to control for retirements, length of service, or other criteria such as ineligibility to reenlist. We merely examine differences between the mobilized/nonmobilized groups in retention by examining whether individuals were there two years later. Nonetheless, the data are interesting and

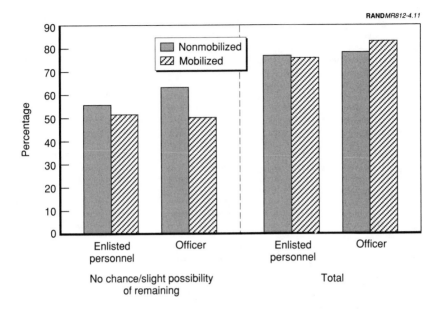

RAND*MR812-4.11*

Figure 4.11—Percentage with Low Probabilities of Remaining in the
Guard/Reserve Still Remaining in the Selected Reserve as of
September 1994: Mobilized and Nonmobilized Reservists

bear further investigation. The effects of mobilization on retention/
reenlistment decisions will be the subject of a future report.

SUMMARY

We find significant differences in the attitudes of spouses and civilian
supervisors of mobilized reservists compared to nonmobilized
reservists.

- A higher proportion of mobilized reserve officers report unfavorable attitudes on the part of both spouses and supervisors
  compared to nonmobilized reservists.

- Mobilized enlisted personnel report a significantly higher
  incidence of unfavorable spouse attitudes but no difference in
  supervisor attitudes. The unfavorable trend in supervisor
  attitudes for officers, but not for enlisted personnel, may reflect

an easier substitutability for jobs performed by enlisted personnel.

- Younger mobilized reservists appear to be at a much higher risk in terms of family conflicts but the shift in attitudes is evident among virtually all ranks.

Overall dissatisfaction with pay and benefits was higher for junior mobilized officer and enlisted personnel than for nonmobilized personnel. However, overall satisfaction with reserve service showed no increased dissatisfaction for officer or enlisted personnel.

There is little evidence here that actual retention decisions have been adversely affected for reservists who were mobilized compared to those not mobilized. Among those with low probabilities of reenlisting/continuing, problems caused by mobilization did not appear to be an important factor in the decision.

Chapter Five

OBSTACLES FACING UNITS IN MEETING TRAINING OBJECTIVES: PERCEPTIONS OF RESERVISTS, 1986 AND 1992

Our new military strategy places increased reliance on the reserves for responding to military contingencies. Therefore, the readiness of reserve units is of crucial importance. The survey data do not allow us to address this question directly; however, we can examine changes in reservists' *perceptions* regarding the ability of their unit to carry out its training objectives. Although the ability to carry out training objectives is only one factor among several that ultimately determine readiness, perceptions of problems facing units in this area are nonetheless important—they can act as early indicators of the need to change personnel or training policies to enhance performance. In addition, perceptions can have important effects on morale and retention and indeed may sometimes prove self-fulfilling.

Respondents to the 1986 and 1992 surveys were asked a series of questions that focused on a number of potential problems that could affect the achievement of unit training and readiness objectives; respondents were then asked to rate these problems along a scale representing the seriousness of the problem. These answers provide a look at the problems facing units as perceived by reserve members today and six years ago.

PERCEIVED PROBLEMS IN MEETING TRAINING OBJECTIVES: ALL COMPONENTS

Tables 5.1 and 5.2 show the overall percentages among the enlisted personnel and officers who regard each particular problem as serious; rankings of the problems are based on these percentages.

Several points can be made from Table 5.1:

- The problems that were ranked high in the 1986 survey by enlisted personnel tend to be ranked high in the 1992 survey, although the rank order has changed a little. In 1986, shortage of time, lack of access to good facilities, and out-of-date equipment/weapons ranked as the top three problems. Despite the Reserve Components' efforts to improve training and access and to modernize, there is still the same level of dissatisfaction regarding these areas in 1992.

  In 1992, these rank as number 3, 2, and 4, respectively, with number 1 being uncertainty about the future status of the unit, a problem that was not included in the 1986 survey for obvious reasons. Reservists are clearly affected by the current uncertain and changing environment; worry about the future status of the unit is mentioned by 16 percent of reservists as a serious problem.

- Perceptions of problems with unit training readiness have on the whole remained remarkably unchanged. The percentages reporting each problem as serious in 1992 are equal to (in some cases, slightly smaller than) those in 1986.

- The majority of reservists do not find many problems with their units. Even the problem that ranked number 1—uncertainty about the future status of the unit—is mentioned by only 16 percent of reservists. Most other problems, particularly personnel-related problems such as low strength, low quality, low attendance, are mentioned as serious by less than one-tenth of the reservists. Time appears to be a important issue: time for paperwork, time for practice, along with the quality and effectiveness of training.

Table 5.1

**Perceived Problems in Meeting Unit Training Objectives:
Rankings by Enlisted Personnel, 1986 and 1992**

Problem	1986 Survey Percent Seeing a Serious Problem	1986 Survey Ranking	1992 Survey Percent Seeing a Serious Problem	1992 Survey Ranking
Not enough time to plan training objectives and get all adminis-trative paperwork done	15.5	1	14.1	3
Lack of access to good training fa-cilities and grounds	15.1	2	15.7	2
Out-of-date equipment/weapons	12.7	3	12.9	4
Not enough drill time to practice skills	9.8	6	9.6	8
Lack of supplies, such as ammuni-tion, gasoline, etc.	11.0	4	10.4	5
Not enough staff resources to plan training	8.2	9	8.9	9
Being below strength in grades E-1 to E-4	10.1	5	10.0	6
Shortage of MOS/Rating/Specialty Qualified personnel	6.9	10	6.4	14
Poor mechanical condition of equipment/weapons	8.5	7	8.3	10
Lack of good instruction manual and materials	8.5	7	10.0	6
Being below strength in grades E-5 to E-9	4.7	14	5.2	17
Low quality of personnel in low grade unit drill positions	5.2	13	5.6	16
Ineffective training during annual training	6.3	11	7.2	12
Low attendance of unit personnel at unit drill	5.4	12	5.8	15
Low attendance of unit personnel at annual training	3.1	15	4.3	18
Excessive turnover of unit personnel	(a)	—	7.6	11
Inability to schedule effective unit annual training, due to gaining command's operating schedule	—	—	6.5	13
Uncertainty about future status of unit	—	—	16.0	1

SOURCE: 1992 Reserve Components Survey, Q.55; 1986 Reserve Components Survey, Q.43.

aThese problems were not included in the 1986 survey.

Table 5.2

Perceived Problems in Meeting Unit Training Objectives: Rankings by Officers, 1986 and 1992

Problem	1986 Survey Percent Seeing a Serious Problem	1986 Survey Ranking	1992 Survey Percent Seeing a Serious Problem	1992 Survey Ranking
Not enough time to plan training objectives and get all administrative paperwork done	25.8	1	21.3	1
Lack of access to good training facilities and grounds	12.3	2	10.8	3
Out-of-date equipment/weapons	10.5	3	8.7	5
Not enough drill time to practice skills	8.9	4	9.2	4
Lack of supplies, such as ammunition, gasoline, etc.	8.0	5	8.1	6
Not enough staff resources to plan training	7.0	6	6.7	7
Being below strength in grades E-1 to E-4	6.7	7	5.3	9
Shortage of MOS/Rating/Specialty Qualified personnel	5.6	8	4.0	13
Poor mechanical condition of equipment/weapons	5.3	9	4.4	11
Lack of good instruction manual and materials	4.6	10	5.5	8
Being below strength in grades E-5 to E-9	3.8	11	3.7	14
Low quality of personnel in low grade unit drill positions	3.2	12	2.3	17
Ineffective training during annual training	2.8	13	3.3	15
Low attendance of unit personnel at unit drill	1.8	14	2.4	16
Low attendance of unit personnel at annual training	1.2	15	2.0	18
Excessive turnover of unit personnel	(a)	—	4.8	10
Inability to schedule effective unit annual training, due to gaining command's operating schedule	—	—	4.2	12
Uncertainty about future status of unit	—	—	13.2	2

SOURCE: 1992 Reserve Components Survey, Q.55; 1986 Reserve Components Survey, Q.43.

aThese problems were not included in the 1986 survey.

Officer rankings are given in Table 5.2. We see the same patterns as among the enlisted personnel.

- The proportion reporting each problem as serious has decreased somewhat between 1986 and 1992 (with some small exceptions). For example, 26 percent of the 1986 respondents rated insufficient time to plan and get paperwork done (which ranked as the highest problem) as a serious problem compared to only 21 percent of the 1992 officers.

- Uncertainty about the future status of the unit appears to be as much a concern among officers as among the enlisted personnel. It ranks second in importance.

- The ranking of the problems does not change substantially between 1986 and 1992. Apart from uncertainty about the future status of the unit—a problem that was not an issue in 1986—lack of time and lack of good training facilities and equipment head the list and mirror the concerns of the enlisted personnel.

- As with the enlisted, perceptions of unit problems are not widespread. Apart from the problems ranked 1 and 2, all the others are mentioned by fewer than 10 percent of officers. Thus, units are by and large regarded as fairly ready to meet training and mission objectives.

Figure 5.1 facilitates a comparison between officer and enlisted rankings. It shows the five most serious problems as ranked by enlisted personnel along with the proportions reporting them as serious problems. It also shows the percentages of officers regarding these as serious problems along with their rankings (shown in parentheses in the figure). There is substantial agreement between the two groups, although shortage of time is clearly a much bigger problem for officers than for enlisted personnel. Personnel problems, which plagued the reserve components in the late 1970s and early 1980s, appear to have been largely solved by the mid-1980s—in both surveys, low quality of unit personnel and shortages of junior or senior staff are mentioned by only 5–10 percent of enlisted personnel and 2–5 percent of officers and do not rank in the top five problems.

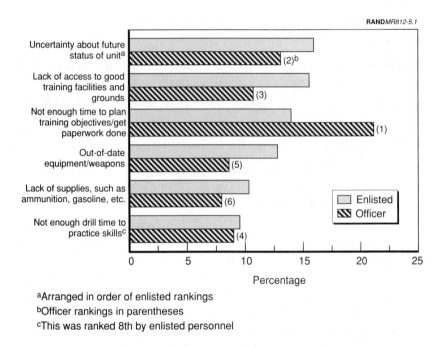

RAND*MR812-5.1*

[a]Arranged in order of enlisted rankings
[b]Officer rankings in parentheses
[c]This was ranked 8th by enlisted personnel

Figure 5.1—Five Most Serious Problems Facing the Unit in Meeting Its Training Objectives, 1992: Enlisted Personnel and Officers

DIFFERENCES AMONG COMPONENTS

We next examine the question of components' facing different sets of problems (as perceived by reservists who serve in those components) and whether particular components appear to be outliers in terms of seriousness of problems when compared to the rest. Tables C.1 and C.2 present the entire set of rankings for both enlisted personnel and officers by component; Tables 5.3 and 5.4 summarize the five most serious problems in each component as ranked by enlisted personnel and officers.

Among the enlisted, reservists in the two Air Reserve components (particularly the ANG) and the Marine Corps Reserve are by far the least troubled about their unit's ability to meet its training objectives; the percentages reporting any given problem as a serious hindrance

Table 5.3

Five Highest-Ranked Problems by Component, 1992: Enlisted Personnel

Problem	ARNG	USAR	NR	MCR	ANG	AFR
Uncertainty about future status of unit	15.8 (1)	19.2 (1)	17.4 (3)	9.8 (3)	8.5 (2)	15.7 (1)
Lack of access to good training facilities and grounds	15.5 (2)	17.9 (2)	22.5 (1)	12.3 (1)	7.2 (4)	9.9 (3)
Below strength in grades E-1 to E-4	14.2 (3)	8.7	10.3	7.7 (5)	3.2	4.6
Out-of-date equipment/weapons	14.0 (4)	15.6 (4)	17.3 (4)	7.6	4.5	4.7
Not enough time to plan training objectives and get all adminis-trative paperwork done	12.4 (5)	16.9 (3)	18.8 (2)	11.7 (2)	8.8 (1)	11.2 (2)
Lack of supplies, such as ammu-nition, gasoline, etc.	9.4	14.0 (5)	14.8 (5)	8.7 (4)	5.0 (5)	3.9
Not enough drill time to practice skills	9.8	10.7	9.8	7.4	7.8 (3)	7.7 (4)
Not enough staff resources to plan training	8.6	11.2	10.8	7.5	4.4	6.6 (5)

SOURCE: 1992 Reserve Components Survey, Q.55, p. 10.

NOTE: Ranks are given in parentheses.

Table 5.4

Five Highest-Ranked Problems by Component, 1992: Officers

Problem	ARNG	USAR	NR	MCR	ANG	AFR
Uncertainty about future status of unit	13.6 (3)	15.3 (2)	11.5 (2)	9.6 (3)	7.9 (2)	12.3 (2)
Lack of access to good training facilities and grounds	15.2 (2)	11.7 (3)	10.2 (3)	7.8 (4)	4.3	5.7 (3)
Below strength in grades E-1 to E-4	11.0	4.5	4.1	5.8 (5)	1.2	2.1
Out-of-date equipment/weapons	12.2 (5)	9.9	7.3 (4)	5.2	5.7 (4)	3.7
Not enough time to plan training objectives and get all adminis-trative paperwork done	20.5 (1)	23.0 (1)	27.5 (1)	19.9 (1)	13.9 (1)	12.8 (1)
Lack of supplies, such as ammu-nition, gasoline, etc.	10.4	10.5 (4)	6.0	4.5	5.1 (5)	2.4
Not enough drill time to practice skills	13.0 (4)	10.4 (5)	5.5	10.8 (2)	6.4 (3)	5.6 (4)
Not enough staff resources to plan training	5.3	8.9	6.8 (5)	5.0	3.3	4.6 (5)

SOURCE: 1992 Reserve Components Survey, Q.55, p. 10.

NOTE: Ranks are given in parentheses.

in meeting unit objectives are very small, generally much less than 10 percent. Naval reservists appear to be the most vocal about problems with 23 percent complaining about lack of access to good training facilities (the highest rate among all the components), and 19 percent complaining about the lack of time for planning and paperwork. Uncertainty about the future and lack of manuals, supplies, and modern equipment/weapons were other major concerns. These were all concerns for the Army Reserve and Army Guard as well (less serious for the Guard in general). The Army Guard also appeared to be concerned with being below strength in the junior grades; this was not a concern for the other components.

Among officers, as with the enlisted, we find that the two Air Reserve components appear to be the most sanguine about their unit's ability to meet training objectives, whereas the Naval Reserve and the two Army components have the highest percentages of officers reporting any given problem as serious. The five highest-ranked concerns mirror those of the enlisted. Time is a serious constraint for all components—it is ranked number 1 by all officers, regardless of component, and it is clearly a far more serious problem for officers than for enlisted. Close to 30 percent of Naval Reserve officers mention it as a serious hindrance as do 20 percent of officers in the ARNG, USAR, and MCR. Uncertainty about the future status of the unit is a serious concern but far fewer in the MCR and ANG mention it (the same pattern as for the enlisted personnel). Other problems frequently mentioned are lack of access to good training facilities (not as serious for the two Air Reserve components) and lack of supplies and modern equipment. In addition, unlike the enlisted personnel, not enough drill time to practice skills is also high on the list of all but the Naval Reserve officers.

SUMMARY

- It is clear that the majority of reservists do not perceive serious problems in their unit's ability to meet training objectives. Overall, there has not been a dramatic shift between 1986 and 1992 in perceptions of the seriousness of problems or the types of problems reservists cite in meeting training objectives. This indicates that peacetime perceptions of units' ability to meet training objectives (and so eventually to carry out their missions) are not changed by the experience of mobilization. Such perceptions may prove useful indicators in designing better training and personnel policies that could enhance training and ultimately readiness.

- It is not surprising that, given the drawdown and the uncertain and changing environment, uncertainty about the future status of the unit (not previously an issue) is a major concern voiced by a significant proportion of both enlisted personnel and officers.

- There is a fair amount of consistency in the problems mentioned by the reservists in the different components, although there is a difference in the perceptions of how serious these problems are. By and large, the air components seem pretty satisfied with their ability to meet training objectives but the naval reserve and the two army components are somewhat less optimistic.

Chapter Six

ABILITY OF UNITS TO MEET TRAINING OBJECTIVES: DIFFERENCES IN PERCEPTIONS OF MOBILIZED AND NONMOBILIZED RESERVISTS

We hypothesized that perceptions of problems facing units in terms of their ability to meet training objectives might differ between mobilized and nonmobilized personnel because of the direct experience of mobilization gained from ODS/S. It was not clear, however, whether mobilization would result in increased perceptions of problems or vice versa. If the experience of being called up revealed shortcomings in equipment, weapons, training, or manning, it was likely that mobilized personnel would be more vocal in their dissatisfaction with their unit and its perceived ability to meet training objectives. However, if the unit was mobilized successfully and performed well in ODS/S, then mobilized personnel in these units would be more satisfied and less likely to complain or find serious faults with their unit.

PERCEPTIONS OF UNITS' ABILITY TO MEET TRAINING OBJECTIVES: MOBILIZED AND NONMOBILIZED RESERVISTS

Tables 6.1 and 6.2 present the rankings of problems by nonmobilized and mobilized personnel based on the percentages of nonmobilized and mobilized personnel reporting a given problem as being serious. By and large, the percentages reporting serious problems are remarkably similar for the two groups among both officers and enlisted personnel. Among the enlisted, mobilized personnel are

Table 6.1

Perceived Problems in Meeting Unit Training Objectives:
Rankings by Enlisted Personnel, 1992

Problem	Nonmobilized Personnel		Mobilized Personnel	
	Percent Seeing a Serious Problem	Ranking	Percent Seeing a Serious Problem	Ranking
Not enough time to plan training objectives and get all administrative paperwork done	13.7	3	15.2	2
Lack of access to good training facilities and grounds	15.7	2	15.6	1
Out-of-date equipment/weapons	13.3	4	11.6	4
Not enough drill time to practice skills	9.7	8	9.3	7
Lack of supplies, such as ammunition, gasoline, etc.	10.4	5	10.5	5
Not enough staff resources to plan training	8.8	10	9.0	8
Being below strength in grades E-1 to E-4	10.4	5	8.9	9
Shortage of MOS/Rating/Specialty Qualified personnel	6.4	14	6.2	15
Poor mechanical condition of equipment/weapons	8.3	11	8.4	11
Lack of good instruction manual and materials	9.9	7	10.5	5
Being below strength in grades E-5 to E-9	4.8	18	6.4	14
Low quality of personnel in low grade unit drill positions	5.7	16	3.5	18
Ineffective training during annual training	7.3	12	6.9	13
Low attendance of unit personnel at unit drill	5.7	16	5.9	16
Low attendance of unit personnel at annual training	9.2	9	4.4	17
Excessive turnover of unit personnel	7.3	12	8.7	10
Inability to schedule effective unit annual training, due to gaining command's operating schedule	6.3	15	7.3	12
Uncertainty about future status of unit	16.7	1	13.6	3

SOURCE: 1992 Reserve Components Survey, Q.55.

Table 6.2

Perceived Problems in Meeting Unit Training Objectives: Rankings by Officers, 1992

Problem	Nonmobilized Personnel		Mobilized Personnel	
	Percent Seeing a Serious Problem	Ranking	Percent Seeing a Serious Problem	Ranking
Not enough time to plan training objectives and get all administrative paperwork done	21.0	1	22.2	1
Lack of access to good training facilities and grounds	10.7	3	11.1	3
Out-of-date equipment/weapons	8.7	5	8.8	6
Not enough drill time to practice skills	8.9	4	9.9	4
Lack of supplies, such as ammunition, gasoline, etc.	7.8	6	8.9	5
Not enough staff resources to plan training	6.5	7	7.3	7
Being below strength in grades E-1 to E-4	5.6	8	4.4	13
Shortage of MOS/Rating/Specialty Qualified personnel	3.9	13	4.5	12
Poor mechanical condition of equipment/weapons	4.2	11	5.0	9
Lack of good instruction manual and materials	5.4	9	5.8	8
Being below strength in grades E-5 to E-9	3.5	14	4.3	14
Low quality of personnel in low grade unit drill positions	2.3	17	2.3	16
Ineffective training during annual training	3.2	15	3.6	15
Low attendance of unit personnel at unit drill	2.6	16	2.0	17
Low attendance of unit personnel at annual training	2.0	18	1.9	18
Excessive turnover of unit personnel	4.8	10	4.7	11
Inability to schedule effective unit annual training, due to gaining command's operating schedule	4.0	12	4.9	10
Uncertainty about future status of unit	12.9	2	14.1	2

SOURCE: 1992 Reserve Components Survey, Q.55.

slightly more concerned about the future status of the unit, but this is not true among officers. Other differences are very small and not significant.

DIFFERENCES IN PERCEPTIONS AMONG COMPONENTS

We also examined the perceptions of mobilized and nonmobilized personnel in different components to see whether there were significant differences among them in their ratings of their unit's ability to meet training objectives (Tables C.3–C.12). Our hypothesis regarding the likelihood of differences in perceptions of unit problems was not borne out: Overall, we found little or no difference among mobilized and nonmobilized reservists with respect to this issue and this was true among both enlisted personnel and officers.

SUMMARY

- Mobilized reservists do not express higher (or lower) levels of concern about unit training problems nor do they express different *types* of concerns from nonmobilized reservists. The ODS/S experience seems not to have shifted reservists' perceptions about their units' training readiness or the types of problems that hinder meeting training objectives.

- The stability of perceptions of problems facing units in meeting training objectives pre- and post-ODS/S and the similarity of perceptions between mobilized and nonmobilized reservists may indicate that reservists are good judges, even in peacetime, of the difficulties units face in meeting training objectives. Their perceptions may provide a good basis for designing training and personnel policies aimed at enhancing unit performance.

Chapter Seven

POTENTIAL PROBLEMS FACING RESERVISTS IN FUTURE RESERVE MOBILIZATIONS: EVIDENCE FROM ODS/S

Reservists who served in ODS/S were the first to experience a large-scale reserve mobilization since Korea. The 1992 survey attempted to capture this experience to assess potential problems that could be associated with future mobilizations and to understand the need for revised compensation and personnel policies to alleviate these problems.

POTENTIAL PROBLEMS IN THE EVENT OF MOBILIZATION

All Reservists

Figures 7.1 and 7.2 present separate data for enlisted personnel and officers on potential problems that reservists are likely to face if called up. In each figure, the percentages shown are for those responding that the given problem would be a serious or somewhat serious one for them and/or their families if they were mobilized. The order of listed problems differs between the two figures but not substantially so. In fact, the five most serious problems are exactly the same. For both groups, loss of income during call-up is the most serious problem reservists would face: 38 percent of the enlisted and 34 percent of the officers listed this as a serious or somewhat serious problem.

The next most serious concerns are those centered on the family: burdens on spouse, increased family problems, and problems for

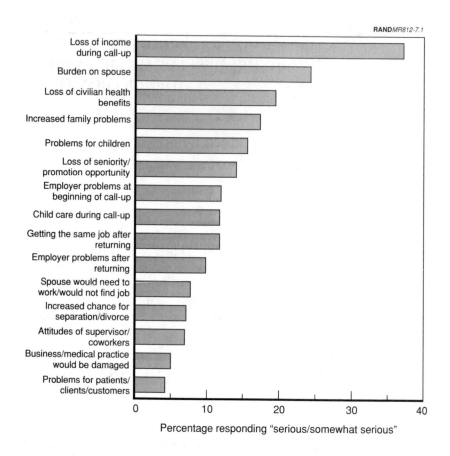

RAND*MR812-7.1*

Figure 7.1—Potential Problems for Enlisted Personnel If Called Up

children. From 25–29 percent are concerned about the burden mobilization would place on their spouses, and another 16–20 percent mentioned more general family problems and problems connected with children.

It is important to note that loss of civilian health benefits ranks quite high for both groups with about one-fifth of the reservists mentioning this as a serious/somewhat serious problem. Preliminary data from the 1991 Survey of Mobilized and Nonmobilized Reservists

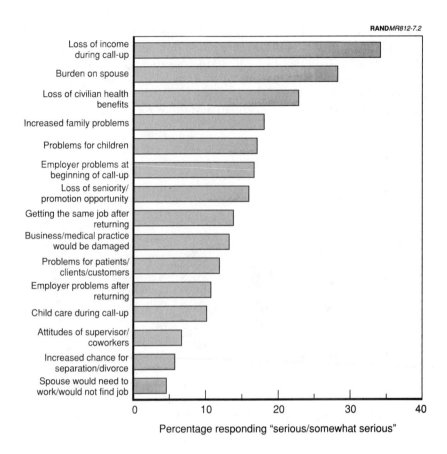

Figure 7.2—Potential Problems for Officers If Called Up

showed that income losses and additional expenses were not the only financial problems suffered by those who were mobilized.

Benefits such as medical, dental, and life insurance, which were normally provided by civilian employers, were either curtailed or terminated for a significant proportion of mobilized reservists. Among those who had an employer-provided benefit—life insurance, medical insurance, dental insurance, pension plan— before mobilization, the proportion of those losing that benefit when mobilized ranged from 40–60 percent, depending on benefit and rank (Figure 7.3).[1] In addition, (not altogether surprising), 77 percent of officers and 81 percent of enlisted personnel did not receive any civilian pay during the period of mobilization. For the other 20 percent, the civilian employer generally replaced some proportion of the difference between their civilian and their military pay.[2]

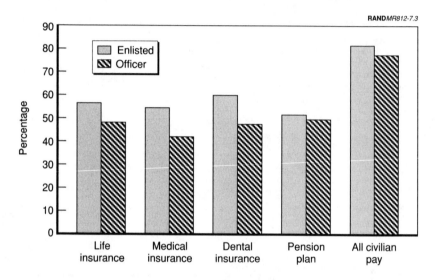

Figure 7.3—Percentage of Mobilized Sample Losing Civilian Benefits and Pay During Mobilization

---

[1]Health and dental benefits should be provided by the military. However, administrative obstacles may prove a formidable barrier to using these benefits.

About 11–25 percent of reservists do not get such benefits from their employers; this group has been omitted from the above calculations. If we include them, we find that about 30–40 percent of all reservists lose civilian benefits of one form or another, when mobilized.

[2]Employers have specific policies regarding civilian pay during the two weeks of annual training, but these policies do not apply to mobilization periods as was clear during ODS/S.

Concern about their civilian jobs ranks next in importance: problems with employers at the beginning of the call-up, loss of seniority or promotion during the time away, and difficulty returning to the same job. In general, attitudes of employers or coworkers did not appear to be a problem. One notable difference between enlisted personnel and officers in the ranking of problems is the higher proportion of officers who responded to questions focusing on business/medical practices and the likely adverse effects of mobilization on such practices. This is due to the higher percentage of self-employed among the officers, particularly doctors.

Differences in Rankings by Employment and Marital Status

The problems faced by individual reservists will depend on their type of employment and marital/family status. To highlight these differences, we examined the answers by both employment status and marital/dependent status. Table 7.1 shows the prevalence of the five most serious problems for enlisted personnel and officers who are self-employed, employed full-time, or employed part-time in the civilian sector. There are some clear-cut differences. Loss of income still dominates the list across the three groups and clearly is more important to those who are self-employed or employed full-time. From 40–45 percent of these groups voice such concerns. Self-employed officers are particularly concerned about the potential damage to businesses or practices and the problems created for clients or patients by their absence—from 50–60 percent of them mention these problems, as do 30–36 percent of the self-employed enlisted. Loss of civilian health benefits mainly affects those who are employed full-time and so are likely to be receiving such benefits.

Table 7.2 shows the rankings of serious problems based on a four-way classification of reservists according to marital status and presence of dependents: single, no dependents; single, with dependents; married, no dependents; married, with dependents. These groups vary with respect to the types of problems they face and the

Table 7.1

Rankings of Most Serious Potential Problems If Called Up
by Employment Status

Problem	Enlisted			Officers		
	Self-Employed	Full-Time	Part-Time	Self-Employed	Full-Time	Part-Time
Loss of income during the call-up	39.4 (1)	43.4 (1)	24.8 (1)	45.0 (3)	36.1 (1)	24.0 (1)
Business/medical practice would be damaged	36.2 (2)	4.6	3.7	59.7 (1)	10.2	11.4
Problems for patients, clients, customers	28.9 (3)	4.0	3.3	51.7 (2)	11.1	9.6
Burden on spouse	25.9 (4)	27.5 (2)	15.1 (2)	29.4 (4)	28.7 (2)	23.9 (2)
Employer problems at the beginning of the call-up	15.8 (5)	13.5	8.6	23.4 (5)	17.6	13.8
Loss of civilian health benefits	10.0	24.3 (3)	9.2	10.8	26.2 (3)	11.6
Increased family problems	16.9	18.2 (4)	14.6 (3)	17.2	18.1 (4)	20.2 (3)
Problems for children	15.3	17.0 (5)	9.9	13.7	16.8	19.0 (4)
Loss of seniority/ promotion opportunity	7.6	16.3	11.3 (4)	7.1	17.9 (5)	11.5
Getting the same job back after returning	7.1	12.9	11.3 (4)	15.0	14.0	8.4
Child care during the call-up	10.1	12.5	8.2	8.3	9.5	14.9 (5)

SOURCE: 1992 Reserve Components Survey, Q.29, p. 7.

resources they have available to meet these problems. Regardless of group, loss of income is the biggest problem reported although, as could be expected, those with dependents appear to be more concerned about the potential loss of income during mobilization. Singles with no dependents face very few family problems and are more concerned with problems related to their civilian job. Burden on spouse is a serious problem for those with spouses, and increased family problems and problems for children are particular concerns of those with dependents. Making provision for child care during call-up is of particular concern to single parents.

Table 7.2

**Rankings of Most Serious Potential Problems If Called Up
by Marital/Dependent Status**

Problem	Single, No Dependents	Single, with Dependents	Married, No Dependents	Married, with Dependents
		Enlisted		
Loss of income during the call-up	30.9	34.5	37.2	42.1
Burden on spouse	—	10.1	31.8	40.3
Loss of civilian health benefits	10.9	16.5	19.6	26.9
Problems for children	—	24.2	—	26.0
Increased family problems	4.5	16.4	17.4	23.2
Child care during the call-up	—	17.6	—	19.6
Loss of seniority/ promotion opportunity	13.8	14.4	14.1	14.1
Employer problems at the beginning of the call-up	12.1	10.6	12.7	12.1
Getting the same job back after returning	12.3	11.4	12.4	11.5
		Officers		
Loss of income during the call-up	27.9	35.4	32.8	35.7
Burden on spouse	—	5.6	26.1	37.8
Loss of civilian health benefits	16.1	25.8	20.2	25.1
Problems for children	—	27.6	—	23.0
Increased family problems	5.2	19.0	15.8	21.9
Child care during the call-up	—	16.5	—	13.9
Loss of seniority/ promotion opportunity	17.0	18.0	18.9	15.0
Employer problems at the beginning of the call-up	17.1	14.7	16.4	17.1
Getting the same job back after returning	14.6	15.1	15.5	13.2

SOURCE: 1992 Reserve Components Survey, Q.29, p. 7.

Differences in Rankings of Potential Problems: Doctors and Pilots

Among officers, we examined data for doctors and pilots separately; recruiting and retention of these groups is a matter of some concern to the Reserve Components and there is both anecdotal and experiential data to suggest that these two groups were hardest hit by ODS/S in terms of income/business losses faced by the reservists during mobilization. Each of these groups constituted a little over 3 percent of the officers. Figure 7.4 shows the rankings of the various

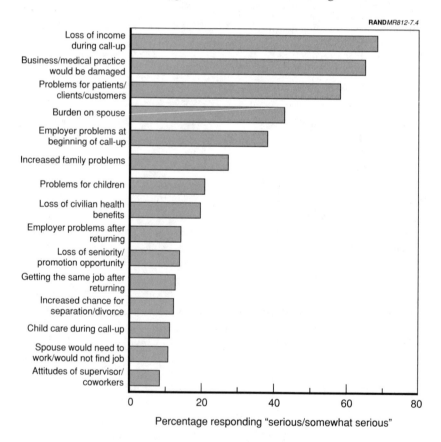

Figure 7.4—Potential Problems for Doctors If Called Up

problems among doctors. Almost 70 percent report that loss of income if mobilized would be a serious/somewhat serious problem—far higher than any of the other groups we have seen thus far. Between 60 and 65 percent are concerned about the effect their absence would have on their medical practices and on their patients; 30 percent are worried about problems caused to their employers (or themselves, if self-employed) when they are first called up; and a fifth are concerned about the loss of civilian benefits. Family issues (with the exception of the burden this would place on their spouses), while important, rank somewhat lower on the scale: about 40 percent report that this would be a burden on their spouse, and 20 percent are worried about increased family problems and problems for children. By and large, doctors appear to be far more concerned about the financial and family burdens of being mobilized than any other group.

It is somewhat surprising that pilots appear to be the opposite of doctors in terms of prevalence and magnitude of problems (Figure 7.5). With the exception of three problems, the percentage mentioning other problems as serious was smaller than 10 percent. Over one-half mentioned loss of income as a serious/somewhat serious problem, another 40 percent were concerned about loss of civilian health benefits (by far, the highest among all the groups), and another 12 percent voiced concern about the possible increased chances for a marital separation/divorce. Other family concerns ranked far lower. One possible explanation is that the nature of a pilot's job itself takes him/her away from home frequently and thus being mobilized and having to be away for extended periods of time is not as likely to cause major family disruptions or problems as for individuals in other occupations. Another reason is that pilots often served in ODS/S for short tours of duty rather than an extended 6-month tour. They were thus less often absent from home or their job.

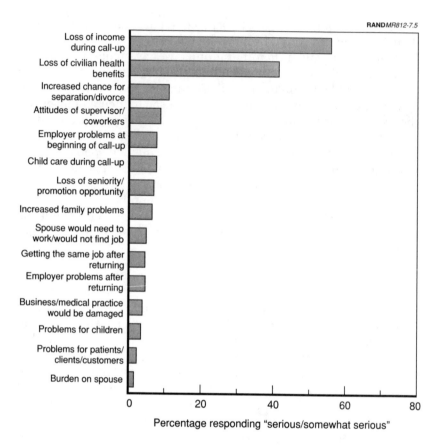

Figure 7.5—Potential Problems for Pilots If Called Up

MOBILIZED VERSUS NONMOBILIZED RESERVISTS

To test whether the perceptions of mobilization problems changed for ODS/S veterans, we analyzed the two groups of mobilized and nonmobilized reservists separately. We first examine differences between the two in terms of reported problems during mobilization. In the case of the mobilized sample, these problems are based on actual experience and we hypothesized that the magnitude of problems for this group was likely to be different (either greater or smaller) than for the nonmobilized reservists. The argument underlying this hypothesis is that mobilization and the kinds of stresses and strains it may bring are unlikely to be fully understood beforehand; however, in some instances, the opposite might be true in that the reality may not prove as bad as the imagining.

Figures 7.6 and 7.7 present the rankings of mobilization problems for mobilized and nonmobilized enlisted personnel and officers. With some exceptions, the type and ranking of problems appear to be very similar for the mobilized and nonmobilized group. Among enlisted personnel, the major differences are that a higher proportion of

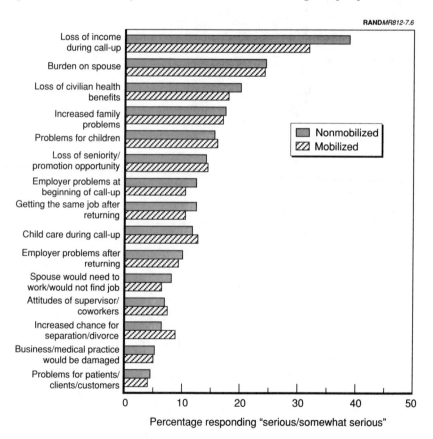

Figure 7.6—Problems for Reservists If Called Up: Mobilized and
Nonmobilized Enlisted Personnel

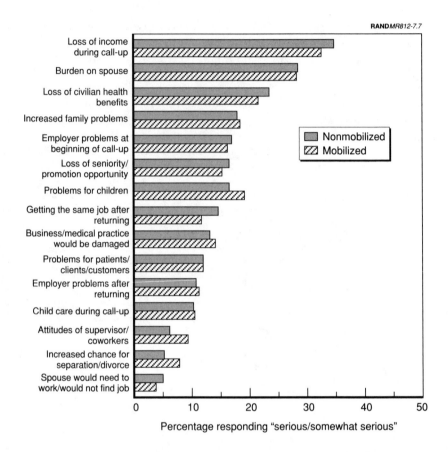

Figure 7.7—Problems for Reservists If Called Up: Mobilized and Nonmobilized Officers

nonmobilized reservists are concerned about income loss compared to mobilized reservists, whereas the latter are more concerned about the increased possibility of divorce/separation.[3] Mobilized reservists are more concerned about the problems for children and the in-

_____

[3]Nonmobilized reservists may find it difficult to measure the difference between what they currently earn (civilian and reserve pay) and what they would earn if called up. Knowing that they would lose most or all civilian pay might lead them to overestimate potential loss in income during mobilization. This may partially account for the difference seen here.

creased chance for separation/divorce than their nonmobilized counterparts. In addition, they appear far more concerned about theattitudes of supervisors and coworkers; this reinforces what we had seen earlier—the increase in unfavorable attitudes towards Guard/Reserve participation reported by mobilized reservists on the part of their civilian supervisors. Although the percentage is still quite low, this is an area to which the reserves may need to pay particular attention, given that our current military plans call for much more frequent use of the reserves.

The data seem to suggest rather strongly that reservists appear to be quite informed about the possible consequences of mobilization and the problems they are likely to face and that the actual experience of being mobilized does not seem very different from their perceptions—with the possible exception of income loss, strains on children and marriages, and some employer problems.

POTENTIAL LOSS IN INCOME DURING MOBILIZATION

The possibility of income loss is the most serious concern for reservists during mobilization—but the threat of income loss appears to be greater than that actually experienced by those mobilized (perhaps for the reason given earlier). Figures 7.8 and 7.9 show the percentage of enlisted personnel and officers reporting that income would decrease greatly if they were to be mobilized for 30 days or more. Overall, 20 percent of the mobilized enlisted sample said they would face a decrease in income compared to 26 percent of the nonmobilized group. This pattern holds true across almost all paygrades. There is little difference among officers, with about 27 percent of both groups reporting a "great" decrease in income if mobilized. The pattern is not clearcut, however: A much higher percentage of O-1s who were mobilized (31 percent) report that their income would decrease greatly relative to those not mobilized (23 percent), but the reverse is true for some of the more senior grades. It may be that the effects of mobilization have been harsher on the more junior officers.

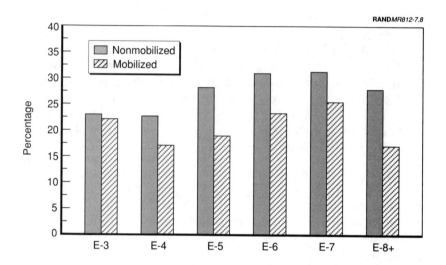

Figure 7.8—Percentage Reporting That Income Would Decrease Greatly If
Mobilized for 30 Days: Mobilized and Nonmobilized Enlisted Personnel

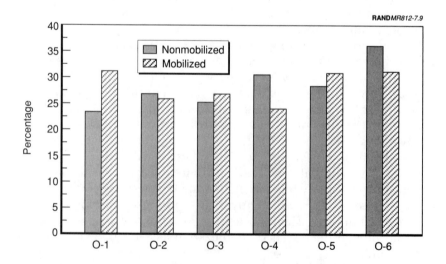

Figure 7.9—Percentage Reporting That Income Would Decrease Greatly If
Mobilized for 30 Days: Mobilized and Nonmobilized Officers

SUMMARY

- Family and economic issues are the dominant concern of reservists when mobilized:

  — The loss of income is the most important concern of reservists, mentioned by 35–40 percent of the reservists;

  — Burden on spouses and increased family problems are mentioned by 20–30 percent; and

  — Loss of civilian health benefits ranks third, mentioned by one-fifth of the reservists.

- Employer-related concerns—problems with employers when mobilized and returning, getting the same job back, loss of good will from clients—are somewhat lower down on the list.

- Potential problems during mobilization vary markedly among different groups of reservists: Self-employed reservists and doctors naturally express very high levels of concern regarding income loss and damage to business or practice; loss of civilian health benefits ranks much higher among pilots than any other group; family problems (burden on spouse, problems for children, etc.) weigh heavily on the minds of those with families.

- Mobilized and nonmobilized reservists do not have drastically different perceptions of problems in a future mobilization, suggesting that reservists have relatively accurate perceptions of the problems they are likely to face if called up, with a couple of exceptions. Nonmobilized reservists are more concerned about income loss and loss of civilian health care benefits compared to mobilized reservists, suggesting that the reality was not as bad as expected; on the other hand, mobilized reservists express more concern about marital stability and concern for children in future mobilizations.

Chapter Eight

CONCLUSIONS

The Selected Reserve has become increasingly senior and more experienced from 1986 to 1992. In addition, the quality of the force has also improved significantly. Despite this, the attitudes, characteristics, and family and work environments of reservists in 1992 are remarkably similar to those reported by reservists in the 1986 survey.[1]

MOTIVATION FOR STAYING IN THE GUARD/RESERVE, FAMILY AND WORK ENVIRONMENTS, 1986 AND 1992

The motivation for staying in the guard/reserve appears to have shifted from 1986 to 1992: Among enlisted personnel, there is less emphasis on immediate compensation and promotion and greater importance placed on educational benefits; among officers, patriotic and job satisfaction motives were more frequently mentioned. Expanded educational benefits may have attracted a newer group of young enlisted personnel whose primary motivation is obtaining money for college rather than long-term reserve service.

There is, however, a small but definite increase in levels of dissatisfaction with military pay and opportunities for education/training

---

[1]We should point out that the 1992 survey has a 50 percent nonresponse rate and that young, single, black, junior reservists tended to have the highest rates of nonresponse. If these reservists have very different attitudes/perceptions/behaviors than their counterparts, then the nonresponse weighting adjustment and subsequent poststratification will not fully compensate for the extent of nonresponse bias (see Appendix A for further details). However, much of our analysis excludes the most junior paygrades (E-1 and E-2), so to that extent, we have avoided the problem of drawing inferences about groups that have very low response rates.

among both officers and enlisted. Part of the dissatisfaction with pay may be a reflection of the perceived higher risk of mobilization and the attendant potential economic losses.

Perhaps the most important positive change is the shift in employer attitudes from 1986 to 1992: a more favorable attitude, fewer conflicts with employers because of the demands of reserve obligations. This may have been partly a result of the significant contributions reservists made during ODS/S.

Perceived attitudes of spouses have remained stable between 1986 and 1992 (a little surprising, given the increased chance of mobilization). These data are particularly important because of the importance of spouse's attitude in reenlistment decisions.

Comparing mobilized with nonmobilized reservists, we find significant differences in the reported attitudes of spouses and civilian supervisors. Higher proportions of mobilized officers reported unfavorable attitudes on the part of both spouses and civilian supervisors than nonmobilized reservists. Among enlisted personnel, we find an increased incidence of unfavorable spouse attitudes among mobilized reservists but little or no difference in supervisor attitudes. Where differences exist, they tend to be much larger among the junior ranks. In addition, junior mobilized officers and enlisted personnel were also much more dissatisfied with pay and benefits than were nonmobilized personnel. However, overall satisfaction with reserve service showed little difference between mobilized and nonmobilized personnel.

RETENTION

This report cannot speak definitively about the effect of mobilization on retention; that is the subject of another report. Nonetheless our data provide a useful first look at retention.

Although reservists across most grades reported much lower subjective probabilities of reenlistment/continuation in the 1992 survey than in the 1986 survey, our simple analysis of continuation rates found little difference between 1986 and 1992. Our data show no dramatic change in overall behavior that could be attributable to

ODS/S. However, we must caution that this finding is based on aggregated, very simple measures.

Our analysis of mobilized and nonmobilized reservists suggests that (a) there is little or no difference in the overall retention rates of these two groups of reservists; and, (b) among officers who expressed serious doubts about continuing, however, mobilized reservists had much lower retention rates than nonmobilized officers.

PERCEIVED PROBLEMS IN MEETING UNIT TRAINING OBJECTIVES, 1986 AND 1992

It is clear that the majority of reservists do not perceive serious problems in their unit's ability to meet training objectives. There is remarkable similarity in the 1986 and 1992 groups in the level and type of concern expressed by reservists about the problems facing units in meeting training objectives. In addition, there is little difference in the rankings of problems by mobilized and nonmobilized reservists. This may indicate that peacetime perceptions of units' ability to meet training objectives could provide a good basis for designing changes in training and personnel policies that could enhance performance. Uncertainty about the future status of the unit (not previously an issue) is a major concern voiced by a significant proportion of both enlisted personnel and officers.

There is a fair amount of consistency in the problems mentioned by the reservists in the different components, although there is a difference in their perceptions of how serious these problems are. By and large, the air components seem pretty satisfied with their ability to meet training objectives but the naval reserve and the two army components are somewhat less optimistic.

POTENTIAL PROBLEMS FOR FUTURE RESERVE MOBILIZATIONS

Family and economic issues dominate the list of problems that reservists are likely to face if mobilized. Potential loss of income is the most important concern of reservists, mentioned by 35–40 percent of the reservists, whereas burden on spouses and increased family problems are mentioned by 20–30 percent. The loss of civilian health benefits ranks third, mentioned by one-fifth of the reservists. Employer-related concerns—problems with employers when mobilized and returning, getting the same job back, damage to business practice, problems for clients and patients—are somewhat lower down on the list.

Potential problems during mobilization vary markedly among different groups of reservists: Self-employed reservists and doctors naturally express very high levels of concern regarding income loss and damage to business or practice; loss of civilian health benefits ranks much higher among pilots than any other group; family problems (burden on spouse, problems for children, etc.) weigh heavily on the minds of those with families. It is important to be aware of these differences when mobilizing specific groups of reservists.

Mobilized and nonmobilized reservists do not have drastically different perceptions of problems in a future mobilization, suggesting that reservists have relatively accurate perceptions of the problems they are likely to face if called up, with a couple of exceptions. Non-mobilized reservists are more concerned about income loss and loss of civilian health care benefits than are mobilized reservists, suggesting that the reality was not as bad as expected; on the other hand, mobilized reservists express more concern about marital stability and concern for children in future mobilizations.

CAVEATS

This report is a simple descriptive analysis of changes in the attitudes, perceptions, and family and work environments of reservists in 1986 and 1992. Part of our focus in this report was on differences between mobilized and nonmobilized reservists in 1992; this is an important addition, given that these provide a first look at how mobilization affects family life, work environments, and the attitudes and behavior of those who were called up. Although the news is largely good, there are two main caveats that must be kept in mind:

- Our data allow us to make inferences regarding the effect of ODS/S on reservists, but we cannot generalize our findings to future mobilizations that might be very different from ODS/S in terms of magnitude, duration, or popularity. The many operations in which reservists have been (and are continuing to be) used since ODS/S need to be carefully studied to gauge the likely effects of different types of mobilizations.[2]

- Current military strategy calls for increased reliance on reserves but it is difficult to assess the effect of increased demands of the reserve job on retention. The increased chance to contribute in meaningful ways to real operations may need to be balanced against the increased likelihood of conflicts with both employers and families. In addition, as the nature and terms of the reserve contract are seen to change, it is difficult to predict the effect on recruiting for similar reasons. Given the importance of the reserves in our military strategy, it is important that manpower planners continue to monitor these indicators and be proactive in forestalling problems in both areas.

POLICY IMPLICATIONS

The chief concern of reservists is potential economic losses if they were mobilized. The Department has considered several options to address this issue including a form of mobilization insurance.

Another major concern is the loss of civilian health care benefits. Although reserve families become eligible for military health care upon mobilization, the loss of civilian benefits can cause serious problems in the continuity and transaction costs associated with medical care. For those losing civilian benefits—a large group, consisting of almost half of those who had such a benefit before mobilization—the burden remains on the family to work out issues of location and access, possible transfer of medical records, and obtaining continuing care for chronic conditions. It would be in the interest of families and probably the reserve forces to direct efforts at maintaining the same civilian-provided health care arrangements for families when reservists are mobilized. This might be done through special CHAMPUS reimbursement mechanisms or through direct payments to families or employers for maintaining civilian health care benefits. However, policies would have to be shaped so that employers who maintain coverage do not shift responsibility to the government.

---

[2]There is a study currently under way at RAND looking at this issue.

Further focus on family-oriented programs for mobilized reservists directed at support of spouses and children seems warranted. The precipitous shift of burdens and responsibilities to spouses upon mobilization, the related effects on children, and possible risk to marital stability might be alleviated through improved access to counseling and support during mobilization, especially during the transition periods, which are likely to be particularly stressful.

Educational benefits were a much more important consideration for younger enlisted personnel in 1992 than in 1986 and may prove an important drawing card for the reserves in the future. These benefits appear to attract higher-quality enlistments who use college benefits while serving in the reserve; the question of whether they are likely to have lower retention is an important, although largely unanswered, one.

1992 SURVEY OF RESERVE PERSONNEL: SAMPLE DESIGN, RESPONSE RATES, WEIGHTS, QUESTIONNAIRE, AND NONRESPONSE

SAMPLE DESIGN AND RESPONSE RATES

The 1992 DoD Reserve Components Survey is one in a series of periodic surveys of officers and enlisted personnel conducted by DoD to collect information regarding the morale, perceptions, and civilian characteristics of reservists. The 1992 survey sample was drawn from the December 1991 reserve population and was updated with current addresses and pay grade as of March 1992.[1] The survey was in the field from October 1992 through late 1993; as a result, the eligible population was redefined to include only those reservists who were in the reserve in both December 1991 and October 1992.

The sample consisted of a random sample drawn from each of four strata:

1. The 1986 longitudinal sample: reservists selected for the 1986 DoD Reserve Components Survey and who were still in the reserves as of December 1991.

2. Individual Mobilization Augmentees (IMAs).

---

[1]The sample excluded those who left or entered the reserve between December 1991 and March 1992.

3. Military technicians.

4. Unit members.

The second, third, and fourth groups were further subdivided into strata based on (a) reserve component, (b) military personnel category (officers versus enlisted personnel), and (c) gender. A simple random sample was selected from each substratum and the sampling rates (defined as the ratio of the sample size to the population) differed across strata. Table A.1 provides information on population and sample sizes and response rates for the four main sampling groups; Table A.2 presents the same information for the unit member sample disaggregated into the various substrata.

Overall, the sampling rate was 0.08 of the total population. The overall response rate was 50 percent, somewhat lower (46 percent) for the unit member sample. For both officers and enlisted personnel, the sampling rates differed by component and gender with women being generally oversampled to provide sufficiently large sample sizes. The response rates were higher among officers than enlisted personnel. However, regardless of military personnel category, the two Air Force Selected Reserve components had the highest response rates among all the components.

Table A.1

Sample Sizes and Response Rates for the 1992 DoD Reserve Components Survey by Main Strata

	December 1991 Population	Sampling Rate	Eligible Sample Size[a]	Response Rate
1986 longitudinal sample	50,849	0.20	9,427	0.57
IMAs	27,966	0.18	4,887	0.61
Military technicians	48,379	0.13	6,007	0.68
Unit members	857,745	0.06	51,758	0.46
All reservists	984,939	0.08	72,079	0.50

SOURCE: Rizzo et al. (1995), Table 1, p. 3.

[a]Sampled individuals had to be members of the Reserve Components in both December 1991 and October 1992 to be eligible to receive a questionnaire. About 6 percent of the original sample of 76,783 were excluded on this basis.

Table A.2

Strata for the Unit Member Officer and Enlisted Personnel Samples: Sample Sizes and Response Rates

Reserve Component	Gender	December 1991 Population	Sampling Rate	Eligible Sample Size[a]	Response Rate
		Officers			
ARNG	Male	31,292	0.07	2,242	0.51
ARNG	Female	2,612	0.15	371	0.49
USAR	Male	29,295	0.07	2,141	0.56
USAR	Female	9,024	0.15	1,306	0.48
NR	Male	19,784	0.08	1,462	0.65
NR	Female	3,688	0.15	541	0.54
MCR	Male	1,983	0.30	589	0.58
MCR	Female	51	0.86	44	0.50
ANG	Male	8,244	0.07	611	0.70
ANG	Female	1,174	0.30	349	0.66
AFR	Male	5,643	0.07	420	0.72
AFR	Female	1,800	0.15	268	0.66
		Enlisted Personnel			
ARNG	Male	285,057	0.05	12,603	0.42
ARNG	Female	19,508	0.10	628	0.36
USAR	Male	151,449	0.05	7,087	0.37
USAR	Female	37,415	0.10	3,475	0.36
NR	Male	75,597	0.05	3,542	0.36
NR	Female	14,788	0.10	1,381	0.38
MCR	Male	33,145	0.10	3,031	0.39
MCR	Female	798	0.50	367	0.34
ANG	Male	58,181	0.05	2,808	0.65
ANG	Female	9,190	0.10	880	0.59
AFR	Male	39,279	0.05	1,903	0.56
AFR	Female	9,365	0.10	917	0.52

SOURCE: Rizzo et al. (1995), Table C-1, pp. C-1–C-2.

[a]Sampled individuals had to be members of the Reserve Components in both December 1991 and October 1992 to be eligible to receive a questionnaire. About 6 percent of the original sample of 76,783 were excluded on this basis.

WEIGHTING THE SURVEY

As Rizzo et al. (1995) point out, the sample design does not produce a self-weighted sample of reservists. As a result, weights are needed to provide unbiased population parameter estimates. These weights accomplish three objectives: (a) They correct for the unequal probabilities of selection of individual reservists, (b) they adjust for survey nonresponse, and (c) they increase the precision of the survey estimates by adjusting the sample proportions to reflect known population characteristics. Rizzo et al. (1995) discuss the three-stage procedure by which survey adjustment weighting was accomplished:

- **Computation of Base Weights.** Base weights are the reciprocal of an individual's selection probability. If 1 in 10 female Air Force officers were selected, the base weight for female Air Force officers would be 10.

- **Adjustment for Nonresponse.** Nonresponse adjustments compensate for the fact that not all sampled individuals returned completed interviews. If 1,000 officers were selected for the sample but only 900 returned completed surveys, the nonresponse adjustment would be 1,000/900 or 1.111.

- **Poststratification to Known Totals.** Poststratification adjusts sample estimates to conform to known population totals. The final stage of survey weighting increases the precision of survey estimates. (p. 4).

A second-stage weight was computed by multiplying the base weight and the nonresponse weighting adjustment. One critical issue in nonresponse weighting adjustment is the construction of the response cells. These cells should be constructed "so as to have differential response rates across cells, and homogeneous response rate within cells. Response adjustment cells should also be related to important attributes and characteristics measured in the survey" (p. 4). The sampling strata: reserve component, military personnel category (officer or enlisted personnel), and gender were used as the response adjustment cells for each of the four sample groups: IMAs, military technicians, unit members, and the 1986 longitudinal group.

The poststratification adjustment was based on two dimensions: The first was based on component, unit membership (unit member, IMAs, military technicians), military personnel category, and gender; the second was based on pay grade category (E-1–E-2, E-3–E-4, E-5–E-6, E-7–E-9, O-1–O-3, O-4+, Warrant Officers) and race (white, black, other). This poststratification adjustment allows the final adjusted weights to produce subcategory totals that mirror those in the reserve master files (see Rizzo et al., 1995 for details).

SURVEY QUESTIONNAIRE

Separate survey instruments were administered to the officers and enlisted personnel (attached as Appendixes E and F). The questionnaire was divided into several broad categories that encompassed military life, civilian employment, and family life; the categories and selected items within each category are shown in Table A.3.

COMPARING RESPONDENTS AND NONRESPONDENTS

To see whether the large nonresponse rate in the 1992 survey may have affected our findings, we present below some selected characteristics of respondents and nonrespondents.

Full Sample

Table A.4 presents the grade distribution of respondents and nonrespondents (unweighted) as well as the weighted distribution. It is clear that the nonrespondents are disproportionately junior: A much higher percentage of junior paygrades did not respond to the survey; this is one of the primary reasons we excluded E-1s and E-2s from the analyses that disaggregated responses by paygrade. Another is that many of them are still very new to military life and unlikely to have enough experience to provide thoughtful comments on reserve problems or policies.

Table A.5 compares the enlisted and officer respondent and nonrespondent samples along a variety of dimensions. Both among the enlisted personnel and officers, we find that nonrespondents tend to have larger proportions of young, single, and black reservists, with a somewhat lower level of educational attainment. They also tend to have fewer years of service. About two-thirds are members of the two Army Reserve components.

It is important to understand what nonresponse weight adjustments do and do not do. These adjustments redress only the "distribution of the sample for the known imbalances, and there can be no guarantee that they will remove—or even reduce—any nonresponse bias. They eliminate nonresponse bias only when the nonrespondents are a random subset of the total sample in each subgroup

Table A.3

Major Topics and Items Included in the 1992 Survey Questionnaire

Topics and Selected Items

I.   Location
      Urban/suburban/central city/rural location
      How long in current neighborhood

II.   Military background
      Reserve component (current and past)
      Pay grade
      Timing of next promotion
      Year of entry into the military
      Years of service
      Reasons for changing unit (for those in a different unit now than two years ago)
      ODS/S mobilization status and how long mobilized

III. Military plans
      Receipt of enlistment/reenlistment bonus
      Likelihood of reenlistment/extension at end of enlisted term of service
      Reasons for not continuing in the reserve at end of enlisted term of service
      Likelihood of staying in the reserve until qualified for retirement
      Concerns about the effect of force reductions
      Preparedness (in terms of wills, power-of-attorney, etc.)
      Problems for family if called up
      Reasons for participating in the reserve

IV.   Military training, benefits, and programs
      Training in current primary specialty (MOS)
      Time spent on working in primary specialty
      Similarity between civilian and military job
      Similarity between reserve and prior active duty specialty
      Attendance at drills and annual training
      Reserve income
      Use of commissary, exchange, and other military facilities
      Eligibility for and use of educational benefits
      Medical/hospitalization coverage/expenses and willingness to purchase
        such coverage through the reserve
      Dental coverage/expenses and willingness to purchase such coverage
        through the reserve
      Problems facing the unit in meeting training objectives
      Transportation to drills
      Satisfaction with training, opportunities to use specialty, promotion,
      leadership
      Rating of weapons and equipment, unit morale, supervision during drills
      Length of time with present unit
      Likelihood of reserve call-up in next five years
      Likelihood of being called up during next mobilization
      Effect on total income in the event of being mobilized
      Duties if mobilized

Table A.3—(continued)

Topics and Selected Items
V.   Individual and family characteristics
Gender, age, citizenship, voting behavior
Race/ethnicity
Educational attainment and aspirations
Marital status and military status, language, age of spouse
Agreement with spouse on civilian and military career plans
Problems for family with respect to attendance at drills and annual training and extra time spent on reserve duties
Spouse's overall attitude toward reserve participation
Number and age of dependents
Family care plan for dependents, elderly relatives
VI.   Civilian work
Current employment, type of employer, hours worked
Problems for employer with respect to attendance at drills and annual training and extra time spent on reserve duties
Time off for reserve duties/civilian pay during reserve duties
Total earned income
Spouse's employment status, income, and hours worked
VII.   Family resources
Income from different sources
Residence: homeowner/renter, total house payment or rent, utilities, maintenance costs, other housing expenses
VIII.   Military life
Feelings about time spent on civilian and reserve jobs, leisure, family, and community activities
Satisfaction with different features of military life and overall satisfaction with time and participation in the reserve

with regard to the survey variables—an unlikely occurrence in practice. Thus, although weighting adjustments may go some way toward compensating for nonresponse, they do not provide a full solution to the problem" (Kalton, 1983, p. 67).

The weights in the 1992 survey do correct for differential nonresponse. However, if nonrespondents differed in attitudes/ perceptions/behavior in significant ways from respondents who formed the basis for the nonresponse adjustment calculations (for example, a larger proportion of nonrespondents were dissatisfied with the reserve and intended to leave, faced greater conflicts with employers, or had very different thoughts about problems facing units in meeting training objectives), then our analyses produce underreports of these various problems/attitudes/behaviors.

Table A.4

Distribution of Respondents and Nonrespondents (Unweighted) and Total (Weighted) by Paygrade, 1992

Characteristic	Respondents (%)	Nonrespondents (%)	Total (%)
Enlisted			
E-1	0.2	1.3	0.5
E-2	2.2	5.7	4.7
E-3	8.5	14.6	10.7
E-4	23.1	33.6	31.8
E-5	29.0	24.8	25.3
E-6	21.2	12.5	17.0
E-7	10.9	5.3	6.9
E-8	3.9	1.6	2.6
E-9	1.0	0.5	0.6
Number	21,259	30,826	717,097
Officer			
O-1	5.7	9.3	7.5
O-2	12.0	18.4	14.6
O-3	29.5	32.8	29.9
O-4	28.1	22.2	25.3
O-5	18.3	13.0	16.7
O-6	5.9	3.8	5.5
O-7	0.3	0.2	0.3
O-8	0.1	0.1	0.2
Number	8,356	6,066	128,667

Mobilized and Nonmobilized Respondents

We are unable to estimate the extent of nonresponse in the mobilized and nonmobilized groups because the categorization variable—mobilization status—was derived from the survey itself. We do not have data on this variable for nonrespondents. However, to examine the differences between mobilized and nonmobilized groups that might account for differences in attitudes and behavior (other than the fact of being mobilized), we examined two sets of comparisons:

• Characteristics of mobilized and nonmobilized respondents in the 1992 survey; and,

Table A.5

Selected Characteristics of Respondents and Nonrespondents (Unweighted), Enlisted Personnel and Officers, 1992

Characteristic	Enlisted Respondents (%)	Enlisted Nonrespondents (%)	Officer Respondents (%)	Officer Nonrespondents (%)
Age				
17–20	7.1	10.0	0.0	.1
21–24	16.6	27.4	1.9	2.9
25–29	17.8	23.4	9.7	17.3
30–34	14.7	13.6	18.4	22.4
35–39	13.5	9.4	23.1	20.3
40–44	14.3	8.3	23.8	18.9
45–49	9.6	4.7	16.3	12.2
50–54	4.1	1.9	5.2	4.2
55+	2.1	1.2	1.7	1.7
Race/ethnicity				
Nonhispanic white	76.4	67.1	89.6	83.1
Black	15.9	23.8	6.0	11.6
Hispanic	4.6	5.6	1.6	1.8
Other	3.1	3.5	2.8	3.5
Marital status				
Married	56.1	44.8	72.7	66.6
Single	43.9	55.2	27.3	33.4
Gender				
Male	79.9	78.4	70.2	65.9
Female	20.1	21.6	29.8	34.1
Education level				
Less than high school	8.2	10.7	0.2	0.1
High school grad.	69.3	73.5	2.3	3.7
Some college	12.1	10.1	8.6	15.6
College graduate	10.4	5.7	88.9	80.6
Years of service				
≤2	13.1	18.6	4.9	8.2
3–5	16.8	27.0	8.1	12.6
6–9	17.4	21.2	14.9	21.3
10–14	19.9	15.6	22.4	21.9
15–19	17.3	8.8	20.8	14.6
20–24	11.9	6.9	19.3	13.8
25+	3.5	1.9	9.6	7.6
Reserve Component				
ARNG	34.6	38.7	15.8	23.7
USAR	23.1	28.0	35.1	41.8
NR	10.2	12.8	16.3	14.0
MCR	7.7	8.6	6.5	5.6
ANG	13.4	5.6	9.3	6.0
AFR	11.0	6.2	17.0	8.9

- Characteristics of mobilized and nonmobilized reservists in the FY91 Selected Reserve inventory; these data allow us to examine indirectly whether the two groups in the 1992 survey had differential nonresponse rates.

Looking first at the 1992 survey respondents, we find that the mobilized and nonmobilized samples are reasonably similar in rank distribution (Table A.6). When we look at other characteristics, we find that the two groups are fairly similar along a number of dimensions (some exceptions: smaller proportion of younger personnel and those with less than 2 years of service (among enlisted); higher proportion of women (among officers) among the mobilized compared to nonmobilized) (Table A.7). However, the components appear to be unevenly represented across the two groups: The mobilized have

Table A.6

Distribution of Mobilized and Nonmobilized Respondents (Unweighted) by Paygrade, 1992

Paygrade	Mobilized (%)	Nonmobilized (%)
Enlisted		
E-1	0.1	0.2
E-2	0.6	2.8
E-3	8.4	8.5
E-4	23.1	23.1
E-5	29.4	28.9
E-6	21.6	21.1
E-7	11.2	10.7
E-8	4.1	3.9
E-9	1.2	0.9
Number	5,365	15,902
Officer		
O-1	2.8	6.8
O-2	12.4	11.8
O-3	31.3	28.9
O-4	27.7	28.3
O-5	19.6	17.8
O-6	6.2	5.8
O-7	0.1	0.4
O-8	0.0	0.2
Number	2,245	6,111

Table A.7

Selected Characteristics of Mobilized and Nonmobilized Respondents (Unweighted), Enlisted Personnel and Officers, 1992

	Enlisted		Officer	
	Mobilized	Nonmobilized	Mobilized	Nonmobilized
Characteristic	(%)	(%)	(%)	(%)
Age				
17–20	4.9	7.9	0.0	0.0
21–24	19.9	15.5	0.6	2.3
25–29	18.6	17.5	7.0	10.7
30–34	14.5	14.8	16.4	19.2
35–39	13.8	13.4	25.8	22.1
40–44	13.8	14.4	26.2	22.9
45–49	8.8	9.9	16.5	16.2
50–54	3.8	4.2	5.0	5.2
55+	1.8	2.3	2.4	1.4
Race/ethnicity				
Nonhispanic white	74.0	77.2	90.0	89.5
Black	18.0	15.2	6.1	5.9
Hispanic	5.1	4.4	1.6	1.6
Other	2.9	3.2	2.3	3.0
Marital status				
Married	53.7	57.0	71.2	73.3
Single	46.3	43.0	28.8	26.7
Gender				
Male	77.9	80.6	62.5	73.1
Female	22.1	19.4	37.5	26.9
Education level				
Less than high school	7.6	8.4	.3	.1
High school grad.	69.8	69.1	2.4	2.3
Some college	12.0	12.1	8.5	8.7
College graduate	10.6	10.3	88.8	88.9
Years of service				
≤2	10.2	14.1	4.2	5.2
3–5	20.5	15.6	8.2	8.1
6–9	18.0	17.2	13.8	15.3
10–14	19.5	20.0	23.1	22.1
15–19	17.7	17.1	23.2	19.9
20–24	11.0	12.2	18.8	19.4
25+	3.1	3.7	8.8	9.9
Reserve Component				
ARNG	23.8	38.3	9.7	18.0
USAR	25.9	22.1	37.2	34.3
NR	6.6	11.4	10.5	18.4
MCR	18.4	4.1	13.8	3.8
ANG	10.3	14.4	8.7	9.5
AFR	15.0	9.6	20.0	15.9

a much higher proportion of MCR, and much lower proportions of NR and ARNG compared to the nonmobilized.

The FY91 inventory can be used to compare the four groups in the 1992 survey file (enlisted/officer mobilized/nonmobilized) (Tables A.8–A.9).

- The 1992 survey mobilized and nonmobilized groups (based on respondents) tend to have a higher rank distribution than the FY91 inventory and this is true for both enlisted personnel and officers. This is because of the large nonresponse among the younger and more junior paygrades seen above.

- Compared to the FY91 inventory of mobilized reservists, the mobilized survey groups have a lower proportion of blacks and males, somewhat higher educational attainment, higher level of experience, and a higher proportion of Air Reserve component personnel. The educational attainment of the respondent

Table A.8

Distribution of Mobilized and Nonmobilized Personnel
(Unweighted) by Paygrade, FY91 Inventory

Characteristic	Mobilized (%)	Nonmobilized (%)
Enlisted		
E-1	0.9	2.7
E-2	2.8	5.2
E-3	11.9	11.0
E-4	28.4	26.2
E-5	27.0	25.5
E-6	17.1	16.8
E-7	8.6	8.9
E-8	2.6	2.8
E-9	0.7	.9
Officer		
O-1	4.9	10.9
O-2	16.3	14.0
O-3	28.0	27.5
O-4	28.0	25.1
O-5	17.3	16.5
O-6	5.2	5.5
O-7	0.1	0.3
O-8	0.0	0.1

Table A.9

Selected Characteristics of Mobilized and Nonmobilized (Unweighted), Enlisted Personnel and Officers, FY91 Inventory

Characteristic	Enlisted		Officer	
	Mobilized (%)	Nonmobilized (%)	Mobilized (%)	Nonmobilized (%)
Age				
17–20	8.3	10.8	0.0	0.1
21–24	24.2	20.6	1.2	3.9
25–29	20.6	19.9	11.5	14.3
30–34	14.1	13.9	16.3	18.6
35–39	11.0	11.2	23.5	20.0
40–44	11.1	11.7	25.5	22.6
45–49	6.3	6.8	13.8	14.4
50–54	2.8	3.2	5.5	4.5
55+	1.6	1.8	2.6	1.6
Race/ethnicity[a]				
Nonhispanic white	70.5	73.6	87.9	87.8
Black	21.5	16.8	6.9	6.7
Hispanic	5.2	6.2	1.9	2.2
Other	2.8	3.5	3.2	3.3
Marital status				
Married	50.0	52.2	71.6	73.1
Single	50.0	47.8	28.4	26.9
Gender				
Male	85.5	88.0	74.4	86.3
Female	14.5	12.0	25.6	13.7
Education level[b]				
Less than high school	10.4	12.6	0.7	0.6
High school grad.	76.4	73.6	10.1	15.8
Some college	7.9	8.1	11.8	13.6
College graduate	5.2	5.6	77.4	69.9
Years of service				
≤2	14.8	18.1	5.4	5.9
3–5	24.4	19.9	9.7	10.4
6–9	19.0	18.9	16.7	16.3
10–14	17.4	17.2	20.5	19.7
15–19	12.2	12.6	21.1	17.5
20–24	8.9	9.7	18.3	19.5
25+	3.3	3.6	8.3	10.8
Reserve Component				
ARNG	28.9	44.2	15.1	30.1
USAR	29.8	22.6	37.3	33.7
NR	8.2	13.7	11.6	18.5
MCR	12.4	2.1	5.5	1.1
ANG	7.1	11.9	8.4	9.1
AFR	13.4	5.3	22.1	7.4

[a]Missing data on 5.0 percent of cases.
[b]Missing data on 8.6 percent of cases.

groups seems remarkably high compared to that of the FY92 inventory; part of this may be because the 1992 survey data are self-reports rather than personnel records; part of it is undoubtedly due to the higher proportion of women and AFR/ANG personnel, all of whom tend to have higher educational levels than other groups.

COSTS AND BENEFITS OF RESERVE PARTICIPATION BY PAYGRADE

Table B.1

Motivation for Reserve Service, 1986 and 1992

Grade	Needed the Money for Basic Family Expenses	Wanted Extra Money to Use Now	Saving Income for the Future
1986 Survey			
Enlisted			
E-3	36.7	38.0	19.8
E-4	37.9	38.2	22.7
E-5	36.8	36.8	21.8
E-6	31.5	32.0	21.4
E-7	24.7	29.1	21.8
E-8	20.3	22.9	21.1
E-9	18.5	22.8	23.4
Total	34.0	34.8	21.9
Officer			
O-1	25.4	24.5	18.9
O-2	25.5	29.1	20.7
O-3	26.2	27.6	21.8
O-4	22.5	26.8	21.7
O-5	15.3	20.1	17.9
O-6	14.7	14.8	12.9
Total	22.6	25.5	20.3
1992 Survey			
Enlisted			
E-3	20.7	25.8	18.4
E-4	26.5	29.8	19.4
E-5	27.2	27.6	20.2
E-6	23.9	24.5	21.4
E-7	22.6	23.2	21.1
E-8	21.1	23.8	22.5
E-9	13.8	16.8	25.1
Total	25.2	27.1	20.2
Officer			
O-1	21.0	20.6	19.6
O-2	20.0	18.9	18.1
O-3	20.8	20.9	19.1
O-4	19.5	19.6	20.3
O-5	15.0	18.2	18.7
O-6	8.1	10.7	15.4
Total	18.4	19.1	19.0

Question: How much have each of the following contributed to your most recent decision to stay in the Guard/Reserve?   (% Major Contribution)

Table B.1 (continued)

Question: How much have each of the following contributed to your most
recent decision to stay in the Guard/Reserve? (% Major Contribution)

Grade	Using Education Benefits	Obtain Training Skills That Would Help Get Civilian Job	Getting Credit Toward Guard/ Reserve Retirement
1986 Survey			
Enlisted			
E-3	24.5	23.5	23.4
E-4	25.3	23.9	34.3
E-5	16.6	16.9	59.0
E-6	9.3	10.9	76.1
E-7	6.9	8.1	83.9
E-8	4.8	6.5	86.8
E-9	1.9	6.1	82.8
Total	16.1	16.4	58.5
Officer			
O-1	28.8	13.8	38.4
O-2	11.2	12.0	47.4
O-3	5.2	4.9	61.9
O-4	2.8	2.8	73.7
O-5	2.7	3.4	77.3
O-6	2.5	2.2	67.6
Total	6.6	5.4	64.6
1992 Survey			
Enlisted			
E-3	44.1	22.7	17.2
E-4	40.8	22.3	32.0
E-5	23.4	16.7	59.1
E-6	15.0	13.2	76.1
E-7	9.6	10.9	83.0
E-8	6.0	7.8	86.5
E-9	5.1	6.6	83.8
Total	27.8	17.7	52.5
Officer			
O-1	30.8	15.2	42.1
O-2	16.4	13.2	41.0
O-3	7.3	6.9	59.0
O-4	3.0	3.6	76.7
O-5	2.3	2.7	78.6
O-6	3.4	2.6	67.6
Total	7.2	6.2	64.7

Table B.1 (continued)

Question: How much have each of the following contributed to your most recent decision to stay in the Guard/Reserve? (% Major Contribution)				
Grade	Serving the Country	Serving with the People in the Unit	Promotion Opportunities	Opportunity to Use Military Equipment
1986 Survey				
Enlisted				
E-3	51.2	29.9	26.3	22.9
E-4	53.7	31.2	30.5	23.3
E-5	54.4	36.0	29.3	18.2
E-6	58.1	41.0	31.5	14.1
E-7	64.0	42.1	35.2	11.8
E-8	67.2	45.7	40.0	10.5
E-9	80.2	48.7	36.6	11.8
Total	56.6	36.6	31.0	17.9
Officer				
O-1	65.1	25.6	33.4	16.9
O-2	58.5	31.2	35.8	14.5
O-3	53.8	28.5	29.2	8.6
O-4	54.8	32.1	33.2	8.6
O-5	63.5	35.8	43.1	7.6
O-6	72.4	42.3	33.8	7.6
Total	58.1	31.6	33.9	9.8
1992 Survey				
Enlisted				
E-3	52.8	18.4	19.8	20.4
E-4	50.6	24.3	21.9	19.1
E-5	58.1	33.9	25.7	18.4
E-6	63.4	40.2	28.8	16.5
E-7	64.2	43.5	32.4	14.9
E-8	72.3	45.9	37.0	14.0
E-9	82.5	50.3	36.8	13.5
Total	57.1	31.7	25.4	18.0
Officer				
O-1	63.6	26.0	34.7	19.4
O-2	66.2	30.6	30.0	14.1
O-3	67.1	31.8	29.2	12.3
O-4	64.8	29.4	30.4	9.9
O-5	68.8	33.1	36.3	7.2
O-6	80.7	39.7	35.4	8.3
Total	67.5	31.6	31.6	11.0

Table B.1 (continued)

Question: How much have each of the following contributed to your most recent decision to stay in the Guard/Reserve? (% Major Contribution)

Grade	Challenge of Military Training	Travel/"Get Away" Opportunities	Just Enjoy the Guard/Reserve	Pride in My Accomplishments in the Guard/Reserve
		1986 Survey		
Enlisted				
E-3	36.1	24.9	23.0	37.9
E-4	36.3	27.3	28.8	44.3
E-5	28.2	28.7	35.0	48.2
E-6	25.9	27.5	40.6	52.7
E-7	27.8	25.6	47.4	61.4
E-8	31.5	23.5	53.6	72.7
E-9	46.4	23.6	68.6	87.5
Total	30.4	27.4	36.3	50.2
Officer				
O-1	39.1	21.4	41.7	54.5
O-2	34.1	22.1	39.7	55.7
O-3	21.8	19.8	36.1	44.8
O-4	17.1	18.7	36.2	45.1
O-5	21.3	17.1	46.1	58.8
O-6	25.9	21.5	56.7	67.3
Total	23.4	19.5	39.6	50.3
		1992 Survey		
Enlisted				
E-3	36.6	23.8	21.1	35.7
E-4	30.9	23.0	23.6	37.6
E-5	27.7	28.1	35.3	48.1
E-6	27.7	27.3	43.3	56.6
E-7	29.7	28.0	48.6	62.6
E-8	31.0	26.4	56.4	72.2
E-9	37.7	27.9	73.8	89.6
Total	29.8	25.8	33.6	47.2
Officer				
O-1	45.1	21.4	42.0	61.1
O-2	36.4	21.2	38.9	56.1
O-3	28.0	20.5	42.1	56.4
O-4	20.6	16.8	42.5	52.3
O-5	18.2	19.1	46.0	56.8
O-6	25.2	17.2	60.7	73.5
Total	25.9	19.2	43.7	56.7

SOURCE: 1992 Reserve Components Survey, Q.30, p. 7; 1986 Reserve Components Survey, Q.26.

Table B.2

Satisfaction with Reserve Pay and Training

Question: All things considered, how satisfied are you with the following?
(Percentages shown are for those reporting "dissatisfied" or "very dissatisfied")

	1986 Survey		1992 Survey	
Grade	Military Pay and Allowances (%)	Opportunities for Education/ Training (%)	Military Pay and Allowances (%)	Opportunities for Education/ Training (%)
Enlisted				
E-3	20.0	24.7	28.8	25.5
E-4	16.4	20.9	20.9	23.9
E-5	13.1	20.3	16.4	21.5
E-6	11.6	19.5	13.1	20.6
E-7	9.0	20.1	9.5	18.4
E-8	8.2	18.0	9.1	16.4
E-9	6.4	11.1	10.1	8.8
Total	13.3	20.2	17.8	22.1
Officer				
O-1	6.5	20.5	7.8	27.0
O-2	8.1	23.1	10.3	30.7
O-3	5.9	20.5	8.2	26.1
O-4	5.0	14.9	6.1	19.8
O-5	4.6	9.8	4.8	14.4
O-6	4.0	7.8	4.4	12.4
Total	5.6	16.7	7.0	22.0

SOURCE: 1992 Reserve Components Survey, Q.144A–H; 1986 Reserve Components Survey, Q.123A–G.

Table B.3

Civilian Pay Status for Annual Training, 1986 and 1992

Grade	Full Civilian Pay + Military (%)	Partial Civilian Pay + Military (%)	Only Military Pay (%)	Served on Days I Didn't Work (%)
		1986 Survey		
Enlisted				
E-3	10.9	8.3	68.9	11.9
E-4	15.5	10.7	62.0	11.8
E-5	26.6	15.7	48.7	9.0
E-6	36.3	18.2	37.9	7.6
E-7	44.2	20.0	28.5	7.3
E-8	46.3	18.1	28.8	6.8
E-9	55.6	16.6	22.7	5.1
Total	28.2	15.3	47.3	9.2
Officer				
O-1	23.9	8.4	53.8	13.9
O-2	34.9	11.1	42.8	11.2
O-3	40.8	13.7	36.5	9.0
O-4	46.1	13.4	29.3	11.2
O-5	52.3	12.6	26.6	8.5
O-6	64.1	8.6	18.7	8.6
Total	43.3	12.5	34.0	10.2
		1992 Survey		
Enlisted				
E-3	9.5	7.7	73.7	9.1
E-4	12.8	10.5	68.3	8.4
E-5	25.6	15.9	51.9	6.6
E-6	36.2	20.2	38.1	5.5
E-7	42.1	21.8	31.0	5.1
E-8	50.7	20.2	24.5	4.6
E-9	55.6	10.5	27.8	6.1
Total	25.6	15.3	52.3	6.9
Officer				
O-1	18.2	15.2	57.4	9.2
O-2	33.0	18.0	42.4	6.6
O-3	38.7	17.9	35.5	7.9
O-4	42.1	19.5	28.7	9.7
O-5	49.2	18.0	24.6	8.1
O-6	56.7	17.8	19.8	5.7
Total	41.3	18.3	32.3	8.2

Question: Which of the following describes how you were paid for the time you took from your civilian job for Guard/Reserve obligations?

SOURCE: 1992 Reserve Components Survey, Q.120; 1986 Reserve Components Survey, Q.107.

Table B.4

Lost Overtime Opportunity for Reservists, 1986 and 1992

Grade	Question: In 1985/1991 did you lose opportunities for overtime/extra pay because of your Guard/Reserve obligations?					
	1986 Survey			1992 Survey		
	Fre- quently (%)	Occasion- ally (%)	No (%)	Fre- quently (%)	Occasion- ally (%)	No (%)
Enlisted						
E-3	23.4	33.8	42.8	18.2	25.4	56.4
E-4	16.9	32.6	50.5	15.0	31.5	53.5
E-5	14.4	33.4	52.2	14.6	32.4	53.0
E-6	13.9	32.3	53.8	13.1	33.5	53.5
E-7	12.5	28.9	58.6	13.3	31.6	55.1
E-8	10.2	26.8	63.0	13.3	22.7	64.0
E-9	8.8	21.4	69.8	8.7	23.2	68.1
Total	14.9	32.1	53.0	14.5	31.5	54.1
Officer						
O-1	11.1	29.4	59.5	8.9	23.4	67.8
O-2	8.2	27.3	64.5	10.0	21.2	9.3
O-3	7.9	16.5	75.6	7.1	17.8	75.1
O-4	6.9	14.0	79.1	5.5	14.9	79.6
O-5	6.1	10.0	83.9	4.6	7.9	87.4
O-6	6.9	9.1	84.0	5.5	5.5	89.0
Total	7.5	16.4	76.1	6.5	14.9	78.6

SOURCE: 1992 Reserve Components Survey, Q.118; 1986 Reserve Components Survey, Q.103.

Table B.5

Overtime Pay Rate for Reservists

	1986 Survey			1992 Survey		
	Not Paid Extra	Paid at Regular Pay Rate	Paid at Higher Rate	Not Paid Extra	Paid at Regular Pay Rate	Paid at Higher Rate
Grade	(%)	(%)	(%)	(%)	(%)	(%)
Enlisted						
E-3	17.9	17.2	64.9	19.5	15.3	65.2
E-4	19.4	16.5	64.1	19.1	14.0	66.9
E-5	23.7	12.6	63.7	24.1	11.4	64.5
E-6	31.5	10.2	58.3	30.0	9.6	60.4
E-7	39.0	8.7	52.4	38.4	9.3	52.4
E-8	46.9	9.3	43.8	47.4	8.8	43.8
E-9	58.8	6.6	34.6	48.8	9.5	41.7
Total	27.0	12.6	60.4	25.5	11.8	62.7
Officer						
O-1	45.7	11.2	43.1	46.8	11.6	41.6
O-2	50.1	14.2	25.7	49.8	9.9	40.3
O-3	68.1	13.2	18.7	65.6	12.1	22.3
O-4	77.7	12.2	10.1	71.5	17.0	11.5
O-5	79.9	14.2	5.9	83.0	10.7	6.3
O-6	82.3	16.0	1.7	84.7	10.7	4.7
Total	70.2	13.2	16.6	69.1	12.7	18.2

Question: In 1985/1991, how were you paid when you worked over 40 hours a week?

SOURCE: 1992 Reserve Components Survey, Q.115; 1986 Reserve Components Survey, Q.102.

Table B.6

Supervisors' Attitudes Toward Reserve Service

	1986 Survey			1992 Survey		
	Very/ Somewhat Favorable	Neither	Somewhat/ Very Un-favorable	Very/ Somewhat Favorable	Neither	Somewhat/ Very Un-favorable
Grade	(%)	(%)	(%)	(%)	(%)	(%)
Enlisted						
E-3	57.3	26.4	16.3	58.1	29.6	12.4
E-4	57.8	27.3	14.9	60.0	28.0	12.0
E-5	57.2	27.3	15.5	61.4	27.4	11.2
E-6	57.8	27.3	14.9	64.0	24.5	11.5
E-7	59.1	26.1	14.8	63.7	24.8	11.5
E-8	62.1	24.1	13.8	66.7	20.6	12.7
E-9	65.1	20.4	14.5	74.9	16.5	8.6
Total	57.9	27.0	15.1	61.7	26.7	11.7
Officer						
O-1	55.0	25.6	19.4	67.0	22.2	10.9
O-2	57.3	24.0	18.7	59.6	27.2	13.2
O-3	55.9	28.8	15.3	62.6	26.0	11.4
O-4	60.3	25.6	14.1	64.2	24.1	11.7
O-5	58.8	29.8	11.4	67.7	21.3	11.1
O-6	71.0	18.5	10.5	72.3	15.9	11.8
Total	58.6	26.6	14.8	64.3	24.0	11.7

Question: What is your immediate (main) civilian supervisor's overall attitude toward your participation in the Guard/Reserve?

SOURCE: 1992 Reserve Components Survey, Q.107; 1986 Reserve Components Survey, Q.94.

Table B.7

Employer-Related Problems Due to Reserve Service

Question: How much of a problem for your main employer (or for you, if you are self-employed) is absence for the following? (Percentages are for those reporting a "serious/somewhat of a problem.")

Grade	Weekend Drills	Annual Training	Extra Time Spent at Guard/Reserve	Time Spent at Work on Guard/ Reserve Business
		1986 Survey		
Enlisted				
E-3	19.0	34.6	31.5	23.3
E-4	15.6	31.3	27.2	19.5
E-5	14.0	27.8	28.3	21.0
E-6	12.7	27.6	25.9	21.0
E-7	12.4	27.6	24.3	21.7
E-8	11.9	27.2	22.1	19.9
E-9	10.4	28.0	23.3	20.6
Total	14.0	28.8	26.8	20.7
Officer				
O-1	15.1	35.6	31.8	25.8
O-2	13.1	30.8	33.5	24.5
O-3	14.6	39.5	33.2	26.8
O-4	13.3	37.6	32.1	26.5
O-5	10.1	35.7	27.0	22.4
O-6	12.8	33.2	25.5	22.1
Total	13.3	36.7	31.4	25.3
		1992 Survey		
Enlisted				
E-3	18.7	28.4	21.4	13.4
E-4	16.6	31.1	21.5	11.9
E-5	12.9	26.0	21.1	12.2
E-6	11.0	23.8	20.4	13.3
E-7	10.7	23.9	19.0	14.1
E-8	9.6	21.6	18.1	17.1
E-9	9.3	21.1	21.4	15.7
Total	13.8	21.1	20.9	12.7
Officer				
O-1	13.6	30.1	27.7	16.7
O-2	11.5	29.6	24.4	14.8
O-3	10.2	29.8	25.9	18.5
O-4	7.1	27.5	20.0	15.4
O-5	8.7	28.5	22.2	17.3
O-6	8.6	24.9	19.8	13.3
Total	9.3	28.6	23.0	16.5

SOURCE: 1992 Reserve Components Survey, Q.108A–D; 1986 Reserve Components Survey, Q.102.

Table B.8

Time Preference for Major Activities

Question: How do you feel about the amount of time you spend on each activity listed below?

	1986 Survey				1992 Survey			
	E-1– E-4	E-5– E-9	O-1– O-3	O-3+	E-1– E-4	E-5– E-9	O-1– O-3	O-3+
Civilian job								
Spend too much time	14.9	20.1	23.5	28.3	18.5	23.1	29.8	23.9
Right amount of time	59.6	68.1	62.5	63.2	58.0	65.6	58.6	59.9
Don't spend enough time	9.2	4.5	5.3	6.0	9.0	3.5	4.3	3.7
Does not apply	16.4	7.4	8.7	2.5	14.6	7.8	7.4	3.5
Total	100.0	100.0	100.0	100.0	100.0	100.0	100.0	100.0
Reserve job								
Spend too much time	9.5	10.0	20.5	26.3	10.8	9.5	17.8	21.8
Right amount of time	78.5	81.4	69.1	68.6	76.1	80.8	70.9	70.5
Don't spend enough time	10.1	7.6	9.6	4.3	9.5	7.8	9.7	6.0
Does not apply	1.8	1.0	0.7	0.3	3.7	1.9	1.6	1.7
Total	100.0	100.0	100.0	100.0	100.0	100.0	100.0	100.0
Family activities								
Spend too much time	1.9	0.6	0.3	0.0	1.1	0.5	0.3	0.2
Right amount of time	38.2	31.1	27.4	23.2	34.3	30.3	26.8	26.9
Don't spend enough time	53.9	64.1	66.7	73.7	58.6	65.0	68.0	70.2
Does not apply	5.9	4.2	5.6	3.0	6.1	4.2	4.9	2.8
Total	100.0	100.0	100.0	100.0	100.0	100.0	100.0	100.0
Leisure Time								
Spend too much time	5.7	1.8	1.3	0.6	3.9	1.9	1.01	0.6
Right amount of time	38.9	31.0	28.5	22.8	35.1	29.1	25.9	23.4
Don't spend enough time	52.3	65.4	69.5	76.1	58.3	67.6	72.4	75.7
Does not apply	3.1	1.8	0.7	0.5	2.7	1.5	0.7	0.3
Total	100.0	100.0	100.0	100.0	100.0	100.0	100.0	100.0
Community activities								
Spend too much time	2.1	2.8	2.6	4.8	1.4	2.2	2.2	4.2
Right amount of time	28.8	30.5	29.7	37.0	27.2	28.9	29.7	35.3
Don't spend enough time	45.3	48.6	54.9	51.0	49.0	51.8	57.4	53.6
Does not apply	23.8	18.2	12.8	7.2	22.4	17.2	10.8	6.9
Total	100.0	100.0	100.0	100.0	100.0	100.0	100.0	100.0

SOURCE: 1992 Reserve Components Survey, Q.142; 1986 Reserve Components Survey, Q.121.

Table B.9

Family Problems Due to Reserve Service

Question: How much of a problem for the family is absence for the following?
(Percentages are for those reporting a "serious/somewhat of a problem.")

	1986 Survey			1992 Survey		
Grade	Weekend Drills	Annual Training	Extra Time Spent at Guard/ Reserve	Weekend Drills	Annual Training	Extra Time Spent at Guard/ Reserve
Enlisted						
E-3	18.7	38.3	30.1	25.0	40.9	24.3
E-4	16.3	30.2	22.1	19.2	33.0	20.0
E-5	13.8	24.1	21.1	14.6	26.6	18.0
E-6	14.9	24.0	22.7	14.0	24.9	18.2
E-7	14.5	22.9	22.2	14.5	23.8	19.7
E-8	19.1	24.7	30.6	15.4	21.8	20.8
E-9	20.8	23.2	33.9	13.8	17.0	20.2
Total	15.1	25.2	22.6	15.9	27.5	19.0
Officer						
O-1	13.6	25.0	29.5	23.3	34.0	24.8
O-2	19.6	29.4	36.4	21.5	33.0	31.0
O-3	27.9	37.6	42.9	22.8	33.5	30.7
O-4	30.0	36.6	40.8	23.6	31.9	27.7
O-5	27.2	32.5	34.8	23.7	30.8	26.9
O-6	23.7	26.5	27.9	18.9	22.4	22.7
Total	26.7	34.3	38.5	22.8	31.5	28.3

SOURCE: 1992 Reserve Components Survey, Q.97A–C; 1986 Reserve Components Survey, Q.87A–C.

Table B.10

Spouse Attitude Toward Reserve Service

Question: What is your spouse's overall attitude toward your participation in the Guard/Reserve?

	1986 Survey			1992 Survey		
Grade	Very/ Somewhat Favorable (%)	Neither (%)	Somewhat/ Very Un- favorable (%)	Very/ Somewhat Favorable (%)	Neither (%)	Somewhat/ Very Un- favorable (%)
Enlisted						
E-3	60.6	15.2	24.2	51.9	30.1	18.1
E-4	67.8	15.9	16.3	64.3	18.1	17.7
E-5	75.2	13.9	10.9	73.3	15.2	11.5
E-6	76.6	12.8	10.6	77.5	12.8	9.7
E-7	79.2	11.2	9.6	78.5	12.6	8.9
E-8	79.9	10.7	9.4	81.6	9.1	9.2
E-9	80.1	7.1	12.8	81.3	9.6	9.1
Total	74.9	13.3	11.8	72.6	15.1	12.3
Officer						
O-1	80.5	11.4	8.1	75.9	8.8	15.3
O-2	75.8	10.0	14.2	74.9	10.6	14.5
O-3	74.2	11.5	14.3	78.8	10.2	11.0
O-4	78.7	9.2	12.1	82.2	8.1	9.8
O-5	83.3	8.6	8.1	83.0	7.8	9.2
O-6	84.2	7.9	7.9	86.3	5.1	8.6
Total	78.4	9.8	11.8	80.8	8.7	10.6

SOURCE: 1992 Reserve Components Survey, Q.98; 1986 Reserve Components Survey, Q.88.

PERCEIVED PROBLEMS IN MEETING UNIT TRAINING OBJECTIVES BY COMPONENT: RANKINGS OF OFFICER AND ENLISTED PERSONNEL, 1992

Table C.1

Perceived Problems in Meeting Unit Training Objectives by Reserve Component: Enlisted Personnel

Problem	ARNG	USAR	NR	MCR	ANG	AFR
Not enough time to plan training objectives and get all administrative paperwork done	12.4	16.9	18.8	11.7	8.8	11.2
Lack of access to good training facilities and grounds	15.5	17.9	22.5	12.3	7.2	9.9
Out-of-date equipment/weapons	14.0	15.6	17.3	7.6	4.5	4.7
Not enough drill time to practice skills	9.8	10.7	9.8	7.4	7.8	7.7
Lack of supplies, such as ammunition, gasoline, etc.	9.4	14.0	14.8	8.7	5.0	3.9
Not enough staff resources to plan training	8.6	11.2	10.8	7.5	4.4	6.6
Being below strength in grades E-1 to E-4	14.2	8.7	10.3	7.7	3.2	4.6
Shortage of MOS/Rating/Specialty Qualified personnel	7.3	7.4	7.4	5.8	1.9	3.3
Poor mechanical condition of equipment/weapons	8.8	9.7	12.0	6.4	2.8	3.9
Lack of good instruction manual and materials	9.9	12.8	13.5	6.9	3.6	5.7
Being below strength in grades E-5 to E-9	5.1	7.0	4.7	3.5	2.4	4.1
Low quality of personnel in low grade unit drill positions	7.0	6.5	5.8	4.7	1.6	2.5
Ineffective training during annual training	7.6	7.6	9.6	7.3	4.3	6.0
Low attendance of unit personnel at unit drill	7.4	8.3	2.9	3.5	1.7	1.9
Low attendance of unit personnel at annual training	5.5	5.8	2.3	2.7	1.5	1.6
Excessive turnover of unit personnel	7.6	10.5	7.0	6.2	2.6	4.9
Inability to schedule effective unit annual training, due to gaining command's operating schedule	6.9	7.4	7.7	6.5	3.6	4.0
Uncertainty about future status of unit	15.8	19.2	17.4	9.8	8.5	15.7

SOURCE: 1992 Reserve Components Survey, Q.55, p. 10.

Table C.2

Perceived Problems in Meeting Unit Training Objectives by Reserve Component: Officers

Problem	ARNG	USAR	NR	MCR	ANG	AFR
Not enough time to plan training objectives and get all adminis- trative paperwork done	20.5	23.0	27.5	19.9	13.9	12.8
Lack of access to good training facilities and grounds	15.2	11.7	10.2	7.8	4.3	5.7
Out-of-date equipment/weapons	12.2	9.9	7.3	5.2	5.7	3.7
Not enough drill time to practice skills	13.0	10.4	5.5	10.8	6.4	5.6
Lack of supplies, such as ammuni- tion, gasoline, etc.	10.4	10.5	6.0	4.5	5.1	2.4
Not enough staff resources to plan training	5.3	8.9	6.8	5.0	3.3	4.6
Being below strength in grades E-1 to E-4	11.0	4.5	4.1	5.8	1.2	2.1
Shortage of MOS/Rating/Specialty Qualified personnel	3.8	5.3	3.9	5.2	2.0	1.7
Poor mechanical condition of equipment/weapons	5.6	5.3	3.9	2.4	2.3	2.2
Lack of good instruction manual and materials	4.4	6.9	6.5	2.3	3.2	3.6
Being below strength in grades E-5 to E-9	2.7	4.7	5.0	4.9	0.7	1.9
Low quality of personnel in low grade unit drill positions	2.4	3.1	2.4	0.7	0.6	1.0
Ineffective training during annual training	3.0	3.8	3.1	1.3	1.6	4.0
Low attendance of unit personnel at unit drill	3.6	3.2	1.4	0.2	1.3	0.8
Low attendance of unit personnel at annual training	3.1	2.6	1.0	0.4	0.7	1.0
Excessive turnover of unit per- sonnel	4.3	6.1	5.9	2.6	1.4	2.4
Inability to schedule effective unit annual training, due to gaining command's operating schedule	4.0	4.8	5.0	3.4	2.3	2.8
Uncertainty about future status of unit	13.6	15.3	11.5	9.6	7.9	12.3

SOURCE: 1992 Reserve Components Survey, Q.55, p. 10.

Table C.3

Perceived Problems in Meeting Unit Training Objectives: Rankings by ARNG Enlisted Personnel, 1992

Problem	Nonmobilized Personnel		Mobilized Personnel	
	Percent Seeing a Serious Problem	Ranking	Percent Seeing a Serious Problem	Ranking
Not enough time to plan training objectives and get all administrative paperwork done	12.5	5	12.0	4
Lack of access to good training facilities and grounds	15.5	2	15.5	1
Out-of-date equipment/weapons	14.4	3	11.9	5
Not enough drill time to practice skills	9.9	6	9.3	8
Lack of supplies, such as ammunition, gasoline, etc.	9.5	8	9.1	10
Not enough staff resources to plan training	8.8	9	7.7	11
Being below strength in grades E-1 to E-4	14.3	4	13.5	3
Shortage of MOS/Rating/Specialty Qualified personnel	7.3	13	7.3	13
Poor mechanical condition of equipment/weapons	8.7	10	9.3	8
Lack of good instruction manual and materials	9.9	6	10.3	6
Being below strength in grades E-5 to E-9	5.0	18	6.0	17
Low quality of personnel in low grade unit drill positions	7.1	15	6.8	15
Ineffective training during annual training	7.8	11	6.7	16
Low attendance of unit personnel at unit drill	7.4	12	7.3	13
Low attendance of unit personnel at annual training	5.5	17	5.5	18
Excessive turnover of unit personnel	7.2	14	9.5	7
Inability to schedule effective unit annual training, due to gaining command's operating schedule	6.8	16	7.6	12
Uncertainty about future status of unit	15.9	1	14.8	2

SOURCE: 1992 Reserve Components Survey, Q.55, p. 10.

Table C.4

Perceived Problems in Meeting Unit Training Objectives: Rankings by USAR Enlisted Personnel, 1992

Problem	Nonmobilized Personnel		Mobilized Personnel	
	Percent Seeing a Serious Problem	Ranking	Percent Seeing a Serious Problem	Ranking
Not enough time to plan training objectives and get all administrative paperwork done	16.5	3	18.1	1
Lack of access to good training facilities and grounds	17.8	2	17.9	2
Out-of-date equipment/weapons	15.8	4	14.9	4
Not enough drill time to practice skills	10.9	8	10.2	9
Lack of supplies, such as ammunition, gasoline, etc.	14.4	5	13.1	5
Not enough staff resources to plan training	11.1	7	11.2	7
Being below strength in grades E-1 to E-4	9.3	11	7.1	15
Shortage of MOS/Rating/Specialty Qualified personnel	7.6	13	7.0	16
Poor mechanical condition of equipment/weapons	9.6	10	10.0	10
Lack of good instruction manual and materials	12.6	6	13.0	6
Being below strength in grades E-5 to E-9	6.6	17	8.1	12
Low quality of personnel in low grade unit drill positions	6.7	16	5.9	17
Ineffective training during annual training	7.6	13	7.6	14
Low attendance of unit personnel at unit drill	8.3	12	8.4	11
Low attendance of unit personnel at annual training	5.8	18	5.7	18
Excessive turnover of unit personnel	10.5	9	10.5	8
Inability to schedule effective unit annual training, due to gaining command's operating schedule	7.2	15	8.1	12
Uncertainty about future status of unit	20.8	1	15.0	3

SOURCE: 1992 Reserve Components Survey, Q.55, p. 10.

Table C.5

**Perceived Problems in Meeting Unit Training Objectives: Rankings by
NR Enlisted Personnel, 1992**

Problem	Nonmobilized Personnel		Mobilized Personnel	
	Percent Seeing a Serious Problem	Ranking	Percent Seeing a Serious Problem	Ranking
Not enough time to plan training objectives and get all administrative paperwork done	17.9	3	24.0	2
Lack of access to good training facilities and grounds	22.1	1	24.9	1
Out-of-date equipment/weapons	17.5	4	16.4	4
Not enough drill time to practice skills	9.6	10	10.5	10
Lack of supplies, such as ammunition, gasoline, etc.	14.4	5	17.3	3
Not enough staff resources to plan training	10.5	8	12.8	6
Being below strength in grades E-1 to E-4	10.0	9	12.2	8
Shortage of MOS/Rating/Specialty Qualified personnel	7.2	13	8.8	13
Poor mechanical condition of equipment/weapons	12.2	7	10.7	9
Lack of good instruction manual and materials	13.2	6	15.5	5
Being below strength in grades E-5 to E-9	4.3	16	7.2	15
Low quality of personnel in low grade unit drill positions	5.7	15	6.6	16
Ineffective training during annual training	9.7	11	8.9	12
Low attendance of unit personnel at unit drill	2.6	17	4.4	17
Low attendance of unit personnel at annual training	2.2	18	3.2	18
Excessive turnover of unit personnel	6.6	14	9.3	11
Inability to schedule effective unit annual training, due to gaining command's operating schedule	7.6	12	8.6	14
Uncertainty about future status of unit	18.3	2	12.6	7

SOURCE: 1992 Reserve Components Survey, Q.55, p. 10.

Table C.6

Perceived Problems in Meeting Unit Training Objectives: Rankings by MCR Enlisted Personnel, 1992

Problem	Nonmobilized Personnel		Mobilized Personnel	
	Percent Seeing a Serious Problem	Ranking	Percent Seeinga Serious Problem	Ranking
Not enough time to plan training objectives and get all adminis-trative paperwork done	12.5	1	11.2	2
Lack of access to good training facilities and grounds	12.2	2	12.3	1
Out-of-date equipment/weapons	8.0	5	7.4	7
Not enough drill time to practice skills	9.4	3	6.0	13
Lack of supplies, such as ammuni-tion, gasoline, etc.	7.8	7	9.3	4
Not enough staff resources to plan training	7.1	8	7.8	6
Being below strength in grades E-1 to E-4	6.4	10	8.7	5
Shortage of MOS/Rating/Specialty Qualified personnel	6.6	9	5.2	14
Poor mechanical condition of equipment/weapons	5.6	13	7.0	10
Lack of good instruction manual and materials	6.3	11	7.4	7
Being below strength in grades E-5 to E-9	2.4	17	4.3	16
Low quality of personnel in low grade unit drill positions	4.0	15	5.2	14
Ineffective training during annual training	8.0	5	6.8	12
Low attendance of unit personnel at unit drill	3.5	16	3.5	17
Low attendance of unit personnel at annual training	2.1	18	3.1	18
Excessive turnover of unit personnel	4.8	14	7.2	9
Inability to schedule effective unit annual training, due to gaining command's operating schedule	5.9	12	6.9	11
Uncertainty about future status of unit	9.3	4	10.1	3

SOURCE: 1992 Reserve Components Survey, Q.55, p. 10.

Table C.7

Perceived Problems in Meeting Unit Training Objectives: Rankings by ANG Enlisted Personnel, 1992

Problem	Nonmobilized Personnel Percent Seeing a Serious Problem	Ranking	Mobilized Personnel Percent Seeing a Serious Problem	Ranking
Not enough time to plan training objectives and get all adminis- trative paperwork done	8.1	2	11.7	1
Lack of access to good training facilities and grounds	6.8	4	9.0	3
Out-of-date equipment/weapons	4.1	8	6.3	6
Not enough drill time to practice skills	7.4	3	9.4	2
Lack of supplies, such as ammuni- tion, gasoline, etc.	4.5	5	7.2	5
Not enough staff resources to plan training	4.3	6	5.2	7
Being below strength in grades E-1 to E-4	3.2	11	3.0	14
Shortage of MOS/Rating/Specialty Qualified personnel	1.8	15	2.6	15
Poor mechanical condition of equipment/weapons	2.7	12	3.4	12
Lack of good instruction manual and materials	3.4	10	4.6	8
Being below strength in grades E-5 to E-9	2.2	14	3.3	13
Low quality of personnel in low grade unit drill positions	1.5	16	1.8	18
Ineffective training during annual training	4.3	6	4.3	9
Low attendance of unit personnel at unit drill	1.5	16	2.5	16
Low attendance of unit personnel at annual training	1.3	18	2.0	17
Excessive turnover of unit personnel	2.4	13	3.5	11
Inability to schedule effective unit annual training, due to gaining command's operating schedule	3.5	9	4.2	10
Uncertainty about future status of unit	8.7	1	7.8	4

SOURCE: 1992 Reserve Components Survey, Q.55, p. 10.

Table C.8

Perceived Problems in Meeting Unit Training Objectives: Rankings by AFR Enlisted Personnel, 1992

Problem	Nonmobilized Personnel		Mobilized Personnel	
	Percent Seeing a Serious Problem	Ranking	Percent Seeing a Serious Problem	Ranking
Not enough time to plan training objectives and get all administrative paperwork done	10.1	3	13.4	1
Lack of access to good training facilities and grounds	10.3	2	9.2	3
Out-of-date equipment/weapons	4.5	9	5.2	9
Not enough drill time to practice skills	7.5	4	8.1	4
Lack of supplies, such as ammunition, gasoline, etc.	3.4	13	4.8	12
Not enough staff resources to plan training	6.8	5	6.0	5
Being below strength in grades E-1 to E-4	4.7	8	4.4	13
Shortage of MOS/Rating/Specialty Qualified personnel	3.4	13	2.9	16
Poor mechanical condition of equipment/weapons	3.8	11	4.3	14
Lack of good instruction manual and materials	5.6	7	5.8	7
Being below strength in grades E-5 to E-9	3.4	13	5.5	8
Low quality of personnel in low grade unit drill positions	2.0	17	3.5	15
Ineffective training during annual training	6.3	6	5.4	9
Low attendance of unit personnel at unit drill	2.2	16	1.4	18
Low attendance of unit personnel at annual training	1.5	18	2.0	17
Excessive turnover of unit personnel	4.4	10	6.0	5
Inability to schedule effective unit annual training, due to gaining command's operating schedule	3.5	12	5.2	11
Uncertainty about future status of unit	16.9	1	13.4	1

SOURCE: 1992 Reserve Components Survey, Q.55, p. 10.

Table C.9

Perceived Problems in Meeting Unit Training Objectives: Rankings by ARNG Officers, 1992

Problem	Nonmobilized Personnel		Mobilized Personnel	
	Percent Seeing a Serious Problem	Ranking	Percent Seeing a Serious Problem	Ranking
Not enough time to plan training objectives and get all administrative paperwork done	20.9	1	18.1	1
Lack of access to good training facilities and grounds	15.6	2	12.7	3
Out-of-date equipment/weapons	12.5	5	10.8	6
Not enough drill time to practice skills	13.1	4	12.3	4
Lack of supplies, such as ammunition, gasoline, etc.	10.3	7	11.3	5
Not enough staff resources to plan training	5.0	9	6.7	8
Being below strength in grades E-1 to E-4	11.4	6	8.9	7
Shortage of MOS/Rating/Specialty Qualified personnel	3.6	14	4.7	11
Poor mechanical condition of equipment/weapons	5.6	8	5.3	9
Lack of good instruction manual and materials	4.4	10	4.5	12
Being below strength in grades E-5 to E-9	2.5	18	3.6	14
Low quality of personnel in low grade unit drill positions	2.7	17	1.1	18
Ineffective training during annual training	3.0	16	3.2	15
Low attendance of unit personnel at unit drill	3.8	13	2.2	16
Low attendance of unit personnel at annual training	3.3	15	1.6	17
Excessive turnover of unit personnel	4.2	11	5.0	10
Inability to schedule effective unit annual training, due to gaining command's operating schedule	3.9	12	4.0	13
Uncertainty about future status of unit	13.3	3	15.5	2

SOURCE: 1992 Reserve Components Survey, Q.55, p. 10.

Table C.10

Perceived Problems in Meeting Unit Training Objectives: Rankings by USAR Officers, 1992

Problem	Nonmobilized Personnel		Mobilized Personnel	
	Percent Seeing a Serious Problem	Ranking	Percent Seeing a Serious Problem	Ranking
Not enough time to plan training objectives and get all administrative paperwork done	22.7	1	23.6	1
Lack of access to good training facilities and grounds	11.5	3	12.1	3
Out-of-date equipment/weapons	9.8	6	10.1	6
Not enough drill time to practice skills	10.2	5	10.8	4
Lack of supplies, such as ammunition, gasoline, etc.	10.5	4	10.3	5
Not enough staff resources to plan training	9.1	7	8.5	7
Being below strength in grades E-1 to E-4	4.9	12	3.5	14
Shortage of MOS/Rating/Specialty Qualified personnel	5.3	10	5.5	11
Poor mechanical condition of equipment/weapons	5.1	11	5.8	10
Lack of good instruction manual and materials	7.2	8	6.4	8
Being below strength in grades E-5 to E-9	4.6	14	5.0	12
Low quality of personnel in low grade unit drill positions	3.2	17	3.0	16
Ineffective training during annual training	3.9	15	3.5	14
Low attendance of unit personnel at unit drill	3.5	16	2.5	17
Low attendance of unit personnel at annual training	2.7	18	2.4	18
Excessive turnover of unit personnel	6.2	9	5.9	9
Inability to schedule effective unit annual training, due to gaining command's operating schedule	4.9	12	4.6	13
Uncertainty about future status of unit	14.8	2	16.5	2

SOURCE: 1992 Reserve Components Survey, Q.55, p. 10.

Table C.11

Perceived Problems in Meeting Unit Training Objectives: Rankings by NR Officers, 1992

Problem	Nonmobilized Personnel		Mobilized Personnel	
	Percent Seeing a Serious Problem	Ranking	Percent Seeing a Serious Problem	Ranking
Not enough time to plan training objectives and get all adminis- trative paperwork done	27.4	1	28.2	1
Lack of access to good training facilities and grounds	9.8	3	12.4	2
Out-of-date equipment/weapons	7.3	4	7.4	8
Not enough drill time to practice skills	5.3	8	7.1	9
Lack of supplies, such as ammuni- tion, gasoline, etc.	5.2	9	10.2	4
Not enough staff resources to plan training	6.6	5	7.7	7
Being below strength in grades E-1 to E-4	3.7	13	6.0	11
Shortage of MOS/Rating/Specialty Qualified personnel	3.7	13	4.8	12
Poor mechanical condition of equipment/weapons	3.8	12	4.7	13
Lack of good instruction manual and materials	6.1	7	9.0	5
Being below strength in grades E-5 to E-9	4.6	10	6.8	10
Low quality of personnel in low grade unit drill positions	2.0	16	4.2	14
Ineffective training during annual training	3.1	15	3.1	16
Low attendance of unit personnel at unit drill	1.4	17	1.4	17
Low attendance of unit personnel at annual training	1.0	18	.9	18
Excessive turnover of unit personnel	6.3	6	3.7	15
Inability to schedule effective unit annual training, due to gaining command's operating schedule	4.3	11	8.7	6
Uncertainty about future status of unit	11.5	2	11.9	3

SOURCE: 1992 Reserve Components Survey, Q.55, p. 10.

Table C.12

Perceived Problems in Meeting Unit Training Objectives: Rankings by
MCR Officers, 1992

Problem	Nonmobilized Personnel		Mobilized Personnel	
	Percent Seeing a Serious Problem	Ranking	Percent Seeing a Serious Problem	Ranking
Not enough time to plan training objectives and get all administrative paperwork done	20.8	1	19.1	1
Lack of access to good training facilities and grounds	8.3	4	7.4	4
Out-of-date equipment/weapons	6.7	5	4.0	7
Not enough drill time to practice skills	9.9	2	11.5	2
Lack of supplies, such as ammunition, gasoline, etc.	5.5	9	3.7	9
Not enough staff resources to plan training	6.7	5	3.7	9
Being below strength in grades E-1 to E-4	6.1	8	5.6	5
Shortage of MOS/Rating/Specialty Qualified personnel	6.7	5	3.9	8
Poor mechanical condition of equipment/weapons	2.0	14	2.8	12
Lack of good instruction manual and materials	3.8	11	1.1	15
Being below strength in grades E-5 to E-9	5.3	10	4.5	6
Low quality of personnel in low grade unit drill positions	.6	17	.8	16
Ineffective training during annual training	1.3	15	1.3	14
Low attendance of unit personnel at unit drill	.3	18	.2	17
Low attendance of unit personnel at annual training	.7	16	.2	17
Excessive turnover of unit personnel	2.9	13	2.4	13
Inability to schedule effective unit annual training, due to gaining command's operating schedule	3.4	12	3.4	11
Uncertainty about future status of unit	9.1	3	10.1	3

SOURCE: 1992 Reserve Components Survey, Q.55, p. 10.

Table C.13

Perceived Problems in Meeting Unit Training Objectives: Rankings by ANG Officers, 1992

	Nonmobilized Personnel		Mobilized Personnel	
Problem	Percent Seeing a Serious Problem	Ranking	Percent Seeing a Serious Problem	Ranking
Not enough time to plan training objectives and get all adminis- trative paperwork done	12.6	1	18.8	1
Lack of access to good training facilities and grounds	3.4	6	7.6	6
Out-of-date equipment/weapons	4.8	4	9.0	4
Not enough drill time to practice skills	5.6	3	9.1	3
Lack of supplies, such as ammuni- tion, gasoline, etc.	4.2	5	8.8	5
Not enough staff resources to plan training	3.1	7	4.2	9
Being below strength in grades E-1 to E-4	1.4	13	.3	18
Shortage of MOS/Rating/Specialty Qualified personnel	1.9	10	2.6	11
Poor mechanical condition of equipment/weapons	1.6	12	4.8	8
Lack of good instruction manual and materials	2.6	8	5.5	7
Being below strength in grades E-5 to E-9	.5	16	1.3	15
Low quality of personnel in low grade unit drill positions	.6	17	.8	16
Ineffective training during annual training	1.4	13	2.3	12
Low attendance of unit personnel at unit drill	1.5	11	.6	17
Low attendance of unit personnel at annual training	.4	18	1.5	14
Excessive turnover of unit personnel	1.2	15	2.1	13
Inability to schedule effective unit annual training, due to gaining command's operating schedule	2.0	9	3.7	10
Uncertainty about future status of unit	7.5	2	9.5	2

SOURCE: 1992 Reserve Components Survey, Q.55, p. 10.

Table C.14

Perceived Problems in Meeting Unit Training Objectives: Rankings by AFR Officers, 1992

Problem	Nonmobilized Personnel		Mobilized Personnel	
	Percent Seeing a Serious Problem	Ranking	Percent Seeing a Serious Problem	Ranking
Not enough time to plan training objectives and get all administrative paperwork done	9.4	2	19.6	1
Lack of access to good training facilities and grounds	4.3	4	8.6	3
Out-of-date equipment/weapons	2.2	8	6.4	6
Not enough drill time to practice skills	4.6	3	7.6	4
Lack of supplies, such as ammunition, gasoline, etc.	1.5	13	4.0	10
Not enough staff resources to plan training	3.5	5	6.9	5
Being below strength in grades E-1 to E-4	1.6	10	3.0	13
Shortage of MOS/Rating/Specialty Qualified personnel	1.4	14	2.2	16
Poor mechanical condition of equipment/weapons	1.6	10	3.6	12
Lack of good instruction manual and materials	3.2	7	4.5	8
Being below strength in grades E-5 to E-9	1.4	14	2.7	14
Low quality of personnel in low grade unit drill positions	.8	16	1.3	18
Ineffective training during annual training	3.3	6	5.4	7
Low attendance of unit personnel at unit drill	.3	17	1.9	17
Low attendance of unit personnel at annual training	.3	17	2.3	15
Excessive turnover of unit personnel	1.6	10	3.9	11
Inability to schedule effective unit annual training, due to gaining command's operating schedule	1.9	9	4.5	8
Uncertainty about future status of unit	12.8	1	11.3	2

SOURCE: 1992 Reserve Components Survey, Q.55, p. 10.

1992 RESERVE COMPONENTS SURVEY OF ENLISTED PERSONNEL

1992 Reserve Components
Survey of Enlisted Personnel

Dear Survey Participant:

Please remove this cover before returning your completed survey. To remove cover, fold back and forth at perforation several times, then tear off. Do <u>not</u> return the cover when returning your completed survey.

* U.S. GOVERNMENT PRINTING OFFICE: 1992-312-525/60003

RCS DD–FM & P (OT) 1852

1992 Reserve Components
Survey of Enlisted Personnel

The National Guard and Reserve Components are conducting a survey of Guard/Reserve personnel. You have been selected to participate in this important survey. Please read the instructions before you begin the questionnaire.

PRIVACY NOTICE

AUTHORITY: 10 U.S.C. 136

PRINCIPAL PURPOSE OR PURPOSES:
Information collected in this survey is used to sample attitudes and/or discern perceptions of social problems observed by the Guard and Reserve Components members and to support additional manpower research activities. This information will assist in the formulation of policies which may be needed to improve the environment for Reserve Components members and families.

ROUTINE USES: None

DISCLOSURE: Your survey instrument will be treated as confidential. All identifiable information will be used only by persons engaged in, and for the purposes of, the survey. It will not be disclosed to others or used for any other purpose. Only group statistics will be reported.

Your participation in the survey is voluntary. Failure to respond to any questions will not result in any penalty. However, your participation is encouraged so that the data will be complete and representative.

OFFICE USE ONLY
- ⬭ PN
- ⬭ NR
- ⬭ RF
- ⬭ NE

INSTRUCTIONS FOR COMPLETING THE SURVEY

- Please use a No. 2 pencil.

USE NO. 2 PENCIL ONLY

- Make heavy black marks that fill the circle for your answer.
- Please do not make stray marks of any kind.

INCORRECT MARKS   CORRECT MARK

- Sometimes you will be asked to "Mark one." When this instruction appears, mark the one best answer.

Example:
In what month are you completing the survey?
- ○ August
- ● September
- ○ October
- ○ November
- ○ December
- ○ January
- ○ February

If your answer is "September," then just mark that one circle.

- Sometimes you will be asked to "Mark all that apply." When this instruction appears, you may mark more than one answer.

Example:
In which components have you served? Mark all that apply.
- ● Active Army (USA)
- ○ Army National Guard (ARNG)
- ● Army Reserve (USAR)
- ○ Active Navy (USN)
- ○ Naval Reserve (USNR)
- ○ Active Air Force (USAF)
- ○ Air National Guard (ANG)
- ○ Air Force Reserve (USAFR)
- ○ Active Marine Corps (USMC)
- ○ Marine Corps Reserve (USMCR)
- ○ Active Coast Guard (USCG)
- ○ Coast Guard Reserve (USCGR)

If your answer is "Active Army (USA)" and "Army Reserve (USAR)," then mark the two circles clearly.

- Answers to some of the questions will be on a SEVEN–POINT SCALE.

Example:
How satisfied are you with the opportunities you have for promotion in your unit?

Very   Very
Dissatisfied   Satisfied

①—②—③—●—⑤—⑥—⑦

If your answer is "VERY DISSATISFIED," you would darken the circle for number 1.

If your answer is "VERY SATISFIED," you would darken the circle for number 7.

If your opinion is somewhere in between, you would darken the circle for number 2 or 3 or 4 or 5 or 6.

- If you are asked to give numbers for your answer, please record as shown below.

Example:
How old were you on your last birthday?

If your answer is 24...
Write the numbers in the boxes, making sure that the last number is always placed in the right-hand box.

Fill in the unused boxes with zeros.

Then darken the circle for the matching number below each box.

Write the number in the boxes ——→

Age Last Birthday

| 2 | 4 |

Then fill in the matching circles ——→

USE NO. 2 PENCIL ONLY

I LOCATION

1. In what month are you completing the survey?
Mark one.

○ August
○ September
○ October
○ November
○ December
○ January
○ February

2. Which of the following best describes the type of place where you are living now? Mark one.

○ In military housing on a base/installation
○ In a large city (over 250,000)
○ In a suburb near a large city
○ In a medium-sized city (50,000-250,000)
○ In a suburb near a medium-sized city
○ In a small city or town (under 50,000)
○ On a farm or ranch
○ In a rural area but not on a farm or ranch

3. How long have you lived in your present neighborhood? Mark one.

○ Less than a year
○ 1-2 years
○ 2-3 years
○ 3-5 years
○ 5 years or more

II MILITARY BACKGROUND

4. Of which Reserve Component are you a member?
Mark one.

○ Army National Guard (ARNG)
○ Army Reserve (USAR)
○ Naval Reserve (USNR)
○ Marine Corps Reserve (USMCR)
○ Air National Guard (ANG)
○ Air Force Reserve (USAFR)
○ Coast Guard Reserve (USCGR)

5. What is your present pay grade? Mark one.

ENLISTED GRADES		OFFICER GRADES	
○ E-1	○ E-6	○ W-1	○ O-1
○ E-2	○ E-7	○ W-2	○ O-2
○ E-3	○ E-8	○ W-3	○ O-3
○ E-4	○ E-9	○ W-4	○ O-4
○ E-5			○ O-5
			○ O-6
			○ O-7 and above

6. When do you expect to get your NEXT PROMOTION to a higher pay grade? Mark one.

○ In less than 3 months
○ 3-6 months from now
○ 7-9 months from now
○ 10-12 months from now
○ 13-18 months from now
○ 19 months to 2 years from now
○ 25 months to 3 years from now
○ More than 3 years from now
○ Does not apply, I don't expect any more promotions

7. Do you expect to receive a commission to Warrant Officer or Officer?

○ I am a Warrant Officer or Officer
○ Yes
○ No

8. In what year did you first enter any branch of the military? (If you first entered in the Active Force, record the year you first entered the Active Force.)

Write the number in the boxes ⟶ 19 [ ][ ] Year

Then fill in the matching circles ⟶

⓪ ⓪
① ①
② ②
③ ③
④ ④
⑤ ⑤
⑥ ⑥
⑦ ⑦
⑧ ⑧
⑨ ⑨

9. When you first entered the military, in which component did you serve? Do not include as active service, service for basic and initial training only.
Mark one.

○ Active Army (USA)
○ Army National Guard (ARNG)
○ Army Reserve (USAR)
○ Active Navy (USN)
○ Naval Reserve (USNR)
○ Active Air Force (USAF)
○ Air National Guard (ANG)
○ Air Force Reserve (USAFR)
○ Active Marine Corps (USMC)
○ Marine Corps Reserve (USMCR)
○ Active Coast Guard (USCG)
○ Coast Guard Reserve (USCGR)

USE NO. 2 PENCIL ONLY

.ich components have you served? Do not include
.9 active service, service for basic and initial training only.
Mark all that apply.

○ Active Army (USA)
○ Army National Guard (ARNG)
○ Army Reserve (USAR)
○ Active Navy (USN)
○ Naval Reserve (USNR)
○ Active Air Force (USAF)
○ Air National Guard (ANG)
○ Air Force Reserve (USAFR)
○ Active Marine Corps (USMC)
○ Marine Corps Reserve (USMCR)
○ Active Coast Guard (USCG)
○ Coast Guard Reserve (USCGR)

11. **In all, to the nearest year, how long have you served in the Guard/Reserve?** Do not include active duty years.

○ Less than 1 year

Years

⓪	⓪
①	①
②	②
③	③
④	④
	⑤
	⑥
	⑦
	⑧
	⑨

12. **In all, to the nearest year, how long did you serve in the Active Force/on active duty?** Do not include your initial active duty training for the Guard/Reserve. Include service as FTS-AGR/TAR.

○ I have never served in the Active Force
○ Less than 1 year

Years

⓪	⓪
①	①
②	②
③	③
④	④
	⑤
	⑥
	⑦
	⑧
	⑨

13. **When you finally leave the Guard/Reserve, how many total years of service do you expect to have?** (Include active duty years.)

Years

⓪	⓪
①	①
②	②
③	③
④	④
	⑤
	⑥
	⑦
	⑧
	⑨

14. **Are you in a different unit now than you were two years ago?** Mark one.

○ I have not been in the Guard/Reserve for two years, GO TO QUESTION 17
○ No, I am in the same unit, GO TO QUESTION 17
○ Yes, in a different unit but in the same component
○ Yes, in a different unit in a different component

15. **Why did you change units?** Mark all that apply.

○ I was offered a promotion
○ Promotion was more likely in a new unit
○ I relocated away from the previous unit
○ I wanted to retrain in a different skill
○ I liked the job better in my new unit
○ I liked the people better in my new unit
○ My old unit was disestablished
○ Other reasons

16. **Did you have to retrain in a new skill when you changed units?**

○ Yes
○ No

17. **Were you mobilized/activated/called-up as a Reservist during Operation Desert Shield/Desert Storm?** Mark all that apply.

○ No, GO TO QUESTION 19
○ Yes, deployed to Persian Gulf area
○ Yes, deployed to other overseas location
○ Yes, deployed in the United States
○ Yes, stayed in my local community

USE NO. 2 PENCIL ONLY

w many months were you mobilized/
activated/called-up?

Number Months

⓪	⓪
①	①
②	②
③	③
④	④
	⑤
	⑥
	⑦
	⑧
	⑨

22. In what month and year will you complete your current term of service (or extension) in the Selected Reserve (ETS)?

A
Month

○ January
○ February
○ March
○ April
○ May
○ June
○ July
○ August
○ September
○ October
○ November
○ December

○ Don't know

B
Year

199

⓪
①
②
③
④
⑤
⑥
⑦
⑧
⑨

III MILITARY PLANS

19. At the time of your enlistment or your **most recent** reenlistment (or extension) in the Guard/Reserve, how many years of Selected Reserve service did you sign up for? Mark one.

○ No set number of years
○ 1 year or less
○ 2 years
○ 3 years
○ 4 years
○ 5 years
○ 6 years
○ 7 years
○ 8 years
○ Don't know

20. At the time of your enlistment or most recent reenlistment, did you receive a bonus? Mark one.

○ No
○ Yes, enlistment or affiliation bonus
○ Yes, reenlistment bonus

21. **If** you were eligible to reenlist this year, would you receive a bonus for reenlisting?

○ Yes
○ No
○ Don't know

23. How likely are you to REENLIST OR EXTEND at the end of your current term of service? Assume that all special pays which you currently receive are still available. Mark one.

○ (0 in 10) No chance
○ (1 in 10) Very slight possibility
○ (2 in 10) Slight possibility
○ (3 in 10) Some possibility
○ (4 in 10) Fair possibility
○ (5 in 10) Fairly good possibility
○ (6 in 10) Good possibility
○ (7 in 10) Probable
○ (8 in 10) Very probable
○ (9 in 10) Almost sure
○ (10 in 10) Certain

USE NO. 2 PENCIL ONLY

24. Below are some reasons people have for DECIDING TO LEAVE the National Guard/Reserve. If you decide to leave the Guard/Reserve at the end of your current term, which of these would be your <u>most important reason</u> for leaving? Which would be your <u>second most important</u> reason for leaving?
(Mark one reason under each column.)

I WOULD LEAVE THE GUARD/RESERVE BECAUSE:	(A) Most Important Reason	(B) Second Most Important Reason
a. I am not eligible to reenlist	O	O
b. I am moving to another area	O	O
c. It is too hard to get to my Guard/Reserve unit	O	O
d. I need the time for my education	O	O
e. My unit drills conflict with my civilian job	O	O
f. My unit drills conflict with my family activities	O	O
g. I want more leisure time	O	O
h. I don't like my unit's training	O	O
i. My unit doesn't have modern equipment for training	O	O
j. I'm bored with unit activities	O	O
k. The pay is too low	O	O
l. Promotions are too slow	O	O
m. I've had too many problems getting paid	O	O
n. Problems caused by mobilization/ activation/deployment	O	O

25. How likely are you to <u>stay</u> in the Guard/Reserve until qualified for retirement? Assume that all special pays which you currently receive are still available. Mark one.

- O (0 in 10) No chance
- O (1 in 10) Very slight possibility
- O (2 in 10) Slight possibility
- O (3 in 10) Some possibility
- O (4 in 10) Fair possibility
- O (5 in 10) Fairly good possibility
- O (6 in 10) Good possibility
- O (7 in 10) Probable
- O (8 in 10) Very probable
- O (9 in 10) Almost sure
- O (10 in 10) Certain

26. Do you plan to elect the Reserve Components Survivor Benefit Plan (SBP) when eligible?

- O Does not apply, I don't plan to remain until 20 years
- O I have already elected to participate
- O I have already elected <u>not</u> to participate
- O Yes, upon receipt of my 20-year letter
- O Yes, when I am 60 years old
- O No
- O Uncertain, I am not aware of the plan at all
- O Uncertain, I don't understand the plan clearly
- O Uncertain, I have not made up my mind

27. How concerned are you about the following as a result of current talk about force reductions in the Guard/Reserve? Mark one for each item.

	Very Greatly Concerned	Greatly Concerned	Moderately Concerned	Somewhat Concerned	Not At All Concerned
a. Your long-term opportunities in the Guard/Reserve	O	O	O	O	O
b. The financial burden on you and/or your family should you have to leave the Guard/Reserve unexpectedly	O	O	O	O	O
c. Impact of my unit closing on my community	O	O	O	O	O

28. The questions below are about your preparedness. Mark one for each item.

	Yes	No	Don't Know	Does Not Apply
a. Do you have a current written will?	O	O	O	O
b. Does anyone currently hold your power-of-attorney?	O	O	O	O
c. Do you have life insurance other than SGLI/VGLI?	O	O	O	O
d. Have you filled out a record of emergency data?	O	O	O	O
e. Does your spouse or next-of-kin know where to find your papers?	O	O	O	O
f. Do you verify/update annually your record of emergency data?	O	O	O	O

USE NO. 2 PENCIL ONLY

29. **If you were to be called up, how much of a problem would each of the following be for you or your family?**
Mark one number for each item.

	A Serious Problem						Not A Problem	Don't Know	Does Not Apply
a. Employer problems at the beginning of the mobilization/activation/call-up	①	②	③	④	⑤	⑥	⑦	○	○
b. Employer problems when you returned to your job	①	②	③	④	⑤	⑥	⑦	○	○
c. Getting the same job back after returning	①	②	③	④	⑤	⑥	⑦	○	○
d. Loss of civilian health benefits during the call-up	①	②	③	④	⑤	⑥	⑦	○	○
e. Loss of seniority, promotion opportunity, or job responsibility on civilian job	①	②	③	④	⑤	⑥	⑦	○	○
f. Loss of income during the call-up	①	②	③	④	⑤	⑥	⑦	○	○
g. Attitudes of supervisor or co-workers upon return	①	②	③	④	⑤	⑥	⑦	○	○
h. Business or medical practice would be damaged	①	②	③	④	⑤	⑥	⑦	○	○
i. Problems for patients, clients, customers	①	②	③	④	⑤	⑥	⑦	○	○
j. Spouse would need work but would not find job	①	②	③	④	⑤	⑥	⑦	○	○
k. Increased family problems	①	②	③	④	⑤	⑥	⑦	○	○
l. Increased chances for a marital separation or divorce	①	②	③	④	⑤	⑥	⑦	○	○
m. Problems for children	①	②	③	④	⑤	⑥	⑦	○	○
n. Burden on spouse	①	②	③	④	⑤	⑥	⑦	○	○
o. Child care during the call-up	①	②	③	④	⑤	⑥	⑦	○	○

30. **People participate in the Guard/Reserve for many reasons. How much have each of the following contributed to your most recent decision to stay in the Guard/Reserve?** Mark one for each item.

	Major Contribution	Moderate Contribution	Minor Contribution	No Contribution
a. Serving the country	○	○	○	○
b. Using educational benefits	○	○	○	○
c. Obtaining training in a skill that would help get a civilian job	○	○	○	○
d. Serving with the people in the unit	○	○	○	○
e. Getting credit toward Guard/Reserve retirement	○	○	○	○
f. Promotion opportunities	○	○	○	○
g. Opportunity to use military equipment	○	○	○	○
h. Challenge of military training	○	○	○	○
i. Needed the money for basic family expenses	○	○	○	○
j. Wanted extra money to use now	○	○	○	○
k. Saving income for the future	○	○	○	○
l. Travel/"get away" opportunities	○	○	○	○
m. Just enjoyed the Guard/Reserve	○	○	○	○
n. Pride in my accomplishments in the Guard/Reserve	○	○	○	○

USE NO. 2 PENCIL ONLY

IV MILITARY TRAINING, BENEFITS, AND PROGRAMS

31. How were you trained for your <u>current</u> Primary Occupational Specialty (MOS/Designator/Rating/AFSC)? Mark <u>all</u> that apply.

○ In a formal service school
○ On-the-job training (OJT) in a civilian job
○ In a formal civilian school
○ On-the-job training (OJT) in the active service
○ On-the-job training (OJT) in a Guard/Reserve unit
○ Correspondence course(s)

32. For all of 1991, what percentage of your Guard/Reserve time was spent working in your Primary Occupational Specialty (MOS/Designator/Rating/AFSC)?

○ None   ○ 25-49%   ○ 75-99%
○ 1-24%   ○ 50-74%   ○ 100% (All)

33. Is your current Primary Occupational Specialty (MOS/Designator/Rating/AFSC) the <u>same</u> one you had while on active duty?

○ Does not apply, I don't have   ○ Yes
    prior active duty service   ○ No

34. How <u>similar</u> is your civilian job to your Guard/Reserve duty?

○ Does not apply, I don't have a civilian job
○ Does not apply, my civilian job is as a Guard/Reserve
    military technician
○ Very similar
○ Similar
○ Somewhat similar
○ Not similar at all

35. In calendar year 1991, which of the following did you participate in/perform? Mark <u>all</u> that apply.

○ Drill weekends
○ Annual Training/ACDUTRA
○ Active duty (other than for training)
○ Active duty for school training
○ Guard/Reserve work at my home or on my civilian job

36. In 1991, how many days of Annual Training/ACDUTRA did you attend? <u>Do not</u> include school unless used to satisfy your Annual Training/ACDUTRA requirement.

○ Did not attend 1991 Annual
    Training/ACDUTRA

Days

[two-column digit grid 0–9]

37. Did you attend the 1991 Annual Training/ACDUTRA a few days at a time, a week or more at a time, or all at once?

○ Did not attend 1991 Annual Training/ACDUTRA
○ A few days at a time, several times over the year
○ A week or more at a time
○ All at once

38. In calendar year 1991, how many paid "Workdays," <u>in</u> <u>addition</u> to any regular drill days and Annual Training/ACDUTRA, did you serve?

○ None

Paid Workdays

[three-column digit grid 0–9]

39. In an average month in 1991, how many <u>unpaid</u> hours did you spend at your drill location (place of regular duty)?

○ None

Unpaid
Hours Per Month

[three-column digit grid 0–9]

40. <u>For all of 1991</u>, what was your total Guard/Reserve income <u>BEFORE taxes and deductions</u>? Include any pay from drills, Annual Training/ACDUTRA, enlistment or affiliation bonuses, and any call-ups or other active duty for training. Please give your best estimate.

• Record the amount in the
  boxes. ──────────▶ $ [   ] .00

• Round to the nearest whole
  dollar.

• Fill in the <u>unused</u> boxes with
  zeros. (For example, if your
  answer is $1,503.75, enter
  01504.)

• Then mark the matching circle
  below <u>each</u> box. ──────▶

Total Guard/
Reserve Income

[five-column digit grid 0–9]

USE NO. 2 PENCIL ONLY

41. In an average month in 1991, how often did you and/or your spouse use each of the following?
Mark one for each item.

TIMES USED IN AVERAGE MONTH

	Not Used	Once	Twice	Three to Five Times	Six Times or More
a. Commissary	○	○	○	○	○
b. Exchange	○	○	○	○	○
c. Other military facilities	○	○	○	○	○

42. Which of the following limit your and/or your spouse's use of the commissary and exchange?
Mark all that apply in each column.

	A. Commissary	B. Exchange
Prices	○	○
Stock	○	○
Hours	○	○
Distance	○	○
Military does not allow more frequent use	○	○

43. Are you now eligible for educational benefits as a result of military service? Mark all that apply.
○ No, GO TO QUESTION 45
○ Yes, State benefits for my Guard/Reserve service
○ Yes, Montgomery GI Bill for Selected Reserve
○ Yes, Active Force benefits (VEAP, GI Bill)
○ Don't know/am not sure

44. Which educational benefits are you now using?
Mark all that apply.
○ None
○ State benefits for Guard/Reserve
○ Montgomery GI Bill for Selected Reserve
○ Active Force benefits (VEAP, GI Bill)

45. Which of the following medical/hospitalization coverages do you have? Mark all that apply.
○ My spouse's active duty military coverage
○ My active duty military coverage
○ Veterans' (VA) coverage
○ My civilian employer's health care plan
○ My spouse's civilian employer's plan
○ Other private coverage
○ None, GO TO QUESTION 47

46. How would you rate the coverage provided by the civilian medical insurance which you have?
○ Does not apply, I do not have civilian medical insurance
○ Excellent
○ Good
○ Fair
○ Poor

47. If it were available through your membership in the Guard or Reserve, would you be interested in purchasing medical insurance?
○ Yes, for myself and my family
○ Yes, for myself only
○ Not sure
○ No, GO TO QUESTION 49

48. If you could buy medical insurance through Guard/Reserve participation, what is the maximum premium cost you would be willing to pay per month?
○ Less than $50 per month
○ $50 per month
○ $100 per month
○ $150 per month
○ $200 per month
○ $250 or more per month

49. How much did you spend on health care services and products (for you and your family) last year? Include CHAMPUS deductions, civilian insurance premiums, and drugs, etc. Do not include dental care.
○ Less than $100
○ $100 to $500
○ $501 to $1,000
○ $1,001 to $1,500
○ $1,501 to $2,500
○ More than $2,500
○ Don't know

50. Which of the following dental coverages do you have?
Mark all that apply.
○ My spouse's active duty military coverage
○ My active duty military coverage
○ Veterans' (VA) coverage
○ My civilian employer's dental plan
○ My spouse's civilian employer's plan
○ Other private coverage
○ None, GO TO QUESTION 52

51. How would you rate the coverage provided by the civilian dental insurance which you have?
○ Does not apply, I do not have civilian dental insurance
○ Excellent
○ Good
○ Fair
○ Poor

52. If it were available through your membership in the Guard or Reserve, would you be interested in purchasing dental insurance?
○ Yes, for myself and my family
○ Yes, for myself only
○ Not sure
○ No, GO TO QUESTION 54

USE NO. 2 PENCIL ONLY

53. If you could buy dental insurance through monthly withholding from your Reserve paycheck, what is the maximum premium cost you would be willing to pay per month?

○ Less than $25 per month
○ $50 per month
○ $100 per month
○ $150 per month
○ $200 per month
○ $250 or more per month

54. How much did you spend for dental treatment (for you and your family) last year? (Include civilian premiums as well as direct payments for treatment.)

○ Less than $100
○ $100 – $200
○ $201 – $300
○ $301 – $500
○ $501 – $800
○ $801 – $1,000
○ $1,001 – $2,000
○ More than $2,000
○ Don't know

55. How much of a problem is each of the following for your unit in <u>meeting your unit's training objectives</u>? Please mark the number which shows your opinion on the lines below. For example, people who feel that an item is <u>Not A Problem</u> would mark 7. People who feel that an item is <u>A Serious Problem</u> would mark 1. Others may have opinions somewhere between 1 and 7. Mark one for each item.

	A Serious Problem						Not A Problem	Don't Know
a. Out-of-date equipment/weapons	①	②	③	④	⑤	⑥	⑦	○
b. Poor mechanical condition of equipment/weapons	①	②	③	④	⑤	⑥	⑦	○
c. Being below strength in <u>Grades E-1 – E-4</u>	①	②	③	④	⑤	⑥	⑦	○
d. Being below strength in <u>Grades E-5 – E-9</u>	①	②	③	④	⑤	⑥	⑦	○
e. Not enough staff resources to plan effective training	①	②	③	④	⑤	⑥	⑦	○
f. Low attendance of unit personnel at <u>Unit Drills</u>	①	②	③	④	⑤	⑥	⑦	○
g. Low attendance of unit personnel at <u>Annual Training/ACDUTRA</u>	①	②	③	④	⑤	⑥	⑦	○
h. Ineffective training during <u>Annual Training/ACDUTRA</u>	①	②	③	④	⑤	⑥	⑦	○
i. Shortage of MOS/Rating/Specialty/AFSC qualified personnel	①	②	③	④	⑤	⑥	⑦	○
j. Low quality of personnel in low grade unit drill positions	①	②	③	④	⑤	⑥	⑦	○
k. Not enough drill time to practice skills	①	②	③	④	⑤	⑥	⑦	○
l. Not enough time to plan training objectives <u>and</u> get all administrative paperwork done	①	②	③	④	⑤	⑥	⑦	○
m. Lack of access to good training facilities and grounds	①	②	③	④	⑤	⑥	⑦	○
n. Lack of good instruction manuals and materials	①	②	③	④	⑤	⑥	⑦	○
o. Lack of supplies, such as ammunition, gasoline, etc.	①	②	③	④	⑤	⑥	⑦	○
p. Excessive turnover of unit personnel	①	②	③	④	⑤	⑥	⑦	○
q. Inability to schedule effective unit annual training due to gaining command's operating schedule	①	②	③	④	⑤	⑥	⑦	○
r. Uncertainty about future status of unit	①	②	③	④	⑤	⑥	⑦	○

PLEASE CHECK: HAVE YOU MARKED A CIRCLE FOR <u>EACH</u> ITEM?

56. How do you usually get to the place of <u>regular</u> military duty or drills? Mark one.

○ Drive myself
○ Driven by spouse
○ Driven by another family member
○ Car pool
○ Civilian air transportation
○ Military air transportation
○ Other public transportation
○ Taxi
○ Walk
○ Other

57. How long does it usually take you to get from home to the place where your unit meets/drills? Mark one.

○ 0-19 minutes
○ 20-39 minutes
○ 40-59 minutes
○ 1-2 hours
○ 2-3 hours
○ 3-6 hours
○ 6 hours or more

USE NO. 2 PENCIL ONLY

FOR QUESTION 58 TO QUESTION 67 BELOW, PLEASE MARK THE NUMBER WHICH SHOWS YOUR OPINION ON THE LINE FOLLOWING EACH QUESTION. For example, people who are Very Satisfied would mark 7. People who are Very Dissatisfied would mark 1. Others may have opinions somewhere between 1 and 7.

58. How satisfied are you with the training received during your unit drills?

Very Dissatisfied — Very Satisfied

①—②—③—④—⑤—⑥—⑦

59. How satisfied are you with the opportunities you have to use your MOS/Designator/Rating/Specialty/AFSC skills during unit drills?

Very Dissatisfied — Very Satisfied

①—②—③—④—⑤—⑥—⑦

60. How satisfied are you with the opportunities you have for promotion in your unit?

Very Dissatisfied — Very Satisfied

①—②—③—④—⑤—⑥—⑦

61. How satisfied are you with your opportunities for leadership in your unit?

Very Dissatisfied — Very Satisfied

①—②—③—④—⑤—⑥—⑦

62. In general, how would you describe the weapons or equipment your unit uses during unit drills?

Out-of-Date — Up-to-Date

①—②—③—④—⑤—⑥—⑦

63. In general, how would you describe the mechanical condition of the weapons and equipment your unit uses during training?

Poor — Excellent

①—②—③—④—⑤—⑥—⑦

64. Overall, how satisfied were you with your unit's activities at 1991 Annual Training/ACDUTRA?

○ Does not apply, I didn't attend 1991 Annual Training/ACDUTRA

Very Dissatisfied — Very Satisfied

①—②—③—④—⑤—⑥—⑦

65. In general, how would you describe the morale of military personnel in your unit?

Morale Is Very Low — Morale Is Very High

①—②—③—④—⑤—⑥—⑦

66. In general, how satisfied are you with the supervision and direction given during unit drills?

Very Dissatisfied — Very Satisfied

①—②—③—④—⑤—⑥—⑦

67. How do you feel about not going to the Persian Gulf area during Operation Desert Storm/Desert Shield?

○ Does not apply, I went to the Persian Gulf area

Very Displeased — Very Pleased

①—②—③—④—⑤—⑥—⑦

68. How long have you been in your present unit?

Years in Present Unit

○ Less than 1 year

⓪⓪
①①
②②
③③
④④
⑤
⑥
⑦
⑧
⑨

69. How likely is it that another conflict requiring a Reserve call-up will occur in the next 5 years?

○ (0 in 10) No chance
○ (1 in 10) Very slight possibility
○ (2 in 10) Slight possibility
○ (3 in 10) Some possibility
○ (4 in 10) Fair possibility
○ (5 in 10) Fairly good possibility
○ (6 in 10) Good possibility
○ (7 in 10) Probable
○ (8 in 10) Very probable
○ (9 in 10) Almost sure
○ (10 in 10) Certain

70. How likely is it that you would be called-up if such a mobilization occurred?

○ (0 in 10) No chance
○ (1 in 10) Very slight possibility
○ (2 in 10) Slight possibility
○ (3 in 10) Some possibility
○ (4 in 10) Fair possibility
○ (5 in 10) Fairly good possibility
○ (6 in 10) Good possibility
○ (7 in 10) Probable
○ (8 in 10) Very probable
○ (9 in 10) Almost sure
○ (10 in 10) Certain

USE NO. 2 PENCIL ONLY

71. If you were mobilized for 30 days or more, would your total income:
○ Increase greatly
○ Increase somewhat
○ Remain the same
○ Decrease somewhat
○ Decrease greatly

72. If mobilized, would you mobilize with your present unit?
○ Yes
○ No
○ Don't know

73. If mobilized, would your military duties be the same as your current duties when attending Annual Training/ACDUTRA?
○ Yes
○ No
○ Don't know

74. Are you Army or Air Force National Guard or Reserve?
○ Yes
○ No, GO TO QUESTION 78

75. Are you a military technician, i.e., a civilian employee of the Army or Air Force National Guard or Reserve?
○ Yes
○ No, GO TO QUESTION 78

76. How long have you been employed as a military technician?

○ Less than 1 year

Years as Technician

⓪	⓪
①	①
②	②
③	③
④	④
	⑤
	⑥
	⑦
	⑧
	⑨

77. Do you drill with the same unit that you work in as a technician?
○ Yes
○ No

V INDIVIDUAL AND FAMILY CHARACTERISTICS

78. Are you male or female?
○ Male
○ Female

79. How old were you on your last birthday?

Age Last Birthday

	⓪
①	①
②	②
③	③
④	④
⑤	⑤
⑥	⑥
	⑦
	⑧
	⑨

80. Where were you born?
○ In the United States
○ Outside the United States to military parents
○ Outside the United States to non-military parents

81. Are you an American citizen?
○ Yes
○ No, resident alien
○ No, not a resident alien

82. Did you vote in the last local election? In the last Presidential election?

A. LAST LOCAL ELECTION
○ Yes, in person at the polls
○ Yes, by absentee ballot
○ No

B. LAST PRESIDENTIAL ELECTION
○ Yes, in person at the polls
○ Yes, by absentee ballot
○ No

83. Are you of Spanish/Hispanic origin or descent?
○ Yes
○ No

84. Are you:
○ American Indian/Alaskan Native
○ Black/Negro/African-American
○ Oriental/Asian/Chinese/Japanese/Korean/Filipino/ Pacific Islander
○ White/Caucasian
○ Other

USE NO. 2 PENCIL ONLY

85. AS OF TODAY, what is the highest school grade or academic degree that you have? DO NOT INCLUDE DEGREES FROM TECHNICAL/TRADE OR VOCATIONAL SCHOOLS. Mark one.

○ Less than 12 years of school (no diploma)
○ GED or other high school equivalency certificate
○ High school diploma
○ Some college, but did not graduate
○ 2-year college degree
○ 4-year college degree (BA/BS)
○ Some graduate school
○ Master's degree (MA/MS)
○ Doctoral degree (PhD/MD/LLB)
○ Other degree not listed above

86. If you are now attending civilian schooling, what kind of school is it? Mark all that apply.

○ Does not apply, I do not attend school
○ Vocational/trade/business or other career training school
○ Junior or community college (2-year)
○ Four-year college or university
○ Graduate/professional school
○ Other

87. What is the highest school grade or academic degree that you think you will complete in the future? Mark one.

○ Does not apply, I don't plan to attend school in the future
○ Less than 12 years of school (no diploma)
○ GED or other high school equivalency certificate
○ High school diploma
○ Some college, but will not graduate
○ 2-year college degree
○ 4-year college degree (BA/BS)
○ Some graduate school
○ Master's degree (MA/MS)
○ Doctoral degree (PhD/MD/LLB)
○ Other degree not listed above

88. Have your parents (or guardians), brothers or sisters (include step-brothers and step-sisters) served in or retired from the military? (Include Guard/Reserve.) Mark all that apply.

	A. Father	B. Mother	C. Brother(s)	D. Sister(s)
Never served	○	○	○	○
Currently serving in the military	○	○	○	○
Served less than 8 years and separated	○	○	○	○
Served more than 8 years (but not retired)	○	○	○	○
Retired from the military	○	○	○	○

89. What is your current marital status? Mark only one answer.

○ Married for the first time
○ Remarried
○ Separated
○ Widowed, GO TO QUESTION 100
○ Divorced, GO TO QUESTION 100
○ Never married, GO TO QUESTION 100

90. Is your spouse currently serving on active duty In the Armed Forces or in the Reserve/Guard?

○ No   ○ Yes, in a Reserve/Guard Component
Yes, on active duty in the:
○ Regular Army   ○ Regular Marine Corps
○ Regular Navy   ○ Regular Air Force
                      ○ Regular Coast Guard

91. Has your current spouse ever served in the U.S. Armed Forces, either on active duty or in the Reserve?

○ No, spouse never served
○ Yes, spouse is retired from Service
○ Yes, spouse is separated from Service
○ Yes, spouse is now in Service

92. How many years have you been married to your current spouse?

Years Married

○ Less than 1 year

⓪	⓪
①	①
②	②
③	③
④	④
	⑤
	⑥
	⑦
	⑧
	⑨

93. How old was your current spouse on her or his last birthday?

Age Last Birthday

⓪	⓪
①	①
②	②
③	③
④	④
⑤	⑤
⑥	⑥
	⑦
	⑧
	⑨

94. Does your spouse speak English as the main language at home?

○ Yes
○ No

USE NO. 2 PENCIL ONLY

FOR QUESTIONS 95 AND 96 PLEASE MARK THE NUMBER WHICH SHOWS YOUR <u>OPINION</u> ON THE LINE FOLLOWING EACH QUESTION.

95. How well do you and your spouse agree on <u>your</u> civilian career plans?

Very Well   Not Well
                                                  At All
① — ② — ③ — ④ — ⑤ — ⑥ — ⑦

96. How well do you and your spouse agree on <u>your</u> military career plans?

Very Well   Not Well
                                                  At All
① — ② — ③ — ④ — ⑤ — ⑥ — ⑦

97. How much of a problem for your family are each of the following? Mark one for each item.

	Serious Problem	Somewhat of a Problem	Slight Problem	Not a Problem	Does Not Apply	Don't Know
a. Absence for weekend drills	○	○	○	○	○	○
b. Absence for Annual Training/ACDUTRA	○	○	○	○	○	○
c. Absence for extra time spent at Guard/Reserve	○	○	○	○	○	○

98. What is your spouse's overall attitude toward your participation in the Guard/Reserve? Mark one.
○ Very favorable
○ Somewhat favorable
○ Neither favorable nor unfavorable
○ Somewhat unfavorable
○ Very unfavorable

99. Has your spouse's support for your decision about staying in the military changed in the past year?
○ Yes, increased
○ No, decreased
○ No, has not changed

EVERYBODY ANSWER:

100. How many dependents do you have in each age group? <u>Do not</u> include yourself or your spouse. For the purpose of this question, a dependent is anyone related to you by blood, marriage, or adoption, and who depends on you for over half his or her support.

○ Does not apply, I have no dependents,
   GO TO QUESTION 104

NUMBER OF DEPENDENTS

Age of dependent	None	1	2	3	4	5 or More
a. Under 1 year	○	○	○	○	○	○
b. 1 year to under 2 years	○	○	○	○	○	○
c. 2-5 years	○	○	○	○	○	○
d. 6-13 years	○	○	○	○	○	○
e. 14-22 years	○	○	○	○	○	○
f. 23-64 years	○	○	○	○	○	○
g. 65 years or over	○	○	○	○	○	○

101. Are arrangements for your dependents <u>who live with you</u> realistically workable for each of the following situations? Mark one for each item.
○ Does not apply, my dependents do not live with me.

	Yes	Probably	No
a. Short-term emergency situation such as a mobilization exercise	○	○	○
b. Long-term situation such as being called-up or mobilized	○	○	○

102. Are any of your dependents physically, emotionally, or intellectually handicapped requiring specialized treatment or care?
○ No
○ Yes, temporarily
○ Yes, permanently

103. If you are a single-parent or a military member married to military member, do you have a military family care plan?
○ Does not apply
○ Yes
○ No

104. Do you have elderly relatives for whom you have responsibility even if they are not your legal dependent(s)?
○ No
○ Yes

105. Does this elderly relative live with you?
○ Does not apply
○ Yes
○ No

USE NO. 2 PENCIL ONLY

VI CIVILIAN WORK

A. YOUR OWN EXPERIENCE

106. Are you currently: Mark all that apply.
- ○ Working full-time as an Army or Air Force Guard/Reserve technician, GO TO QUESTION 109
- ○ Working full-time in a civilian job (not technician)
- ○ Working part-time in a civilian job
- ○ With a civilian job but not at work because of temporary illness, vacation, strike, etc.
- ○ Self-employed in own business
- ○ Unpaid worker (volunteer or in family business)
- ○ Unemployed, laid off, looking for work
- ○ Not looking for work but would like to work
- ○ In school
- ○ Retired
- ○ A homemaker
- ○ Other

107. What is your immediate (main) civilian supervisor's overall attitude toward your participation in the Guard/Reserve? Mark one.
- ○ Does not apply, I am not working at a civilian job, GO TO QUESTION 109
- ○ Does not apply, I am self-employed
- ○ Very favorable
- ○ Somewhat favorable
- ○ Neither favorable nor unfavorable
- ○ Somewhat unfavorable
- ○ Very unfavorable

108. How much of a problem for your main employer (or for you, if self-employed) are each of the following?
Mark one for each item.

	Serious Problem	Somewhat of a Problem	Slight Problem	Not a Problem	Does Not Apply	Don't Know
a. Absence for weekend drills	○	○	○	○	○	○
b. Absence for Annual Training/ACDUTRA	○	○	○	○	○	○
c. Absence for extra time spent at Guard/Reserve	○	○	○	○	○	○
d. Time spent while at civilian work on Guard/Reserve business	○	○	○	○	○	○

THE NEXT QUESTIONS ARE ABOUT YOUR CIVILIAN JOB IN 1991. IF YOU HAD MORE THAN ONE JOB, PLEASE ANSWER THESE QUESTIONS FOR THE JOB WHERE YOU WORKED THE MOST HOURS PER WEEK FOR MOST OF THE YEAR.

109. What kind of work did you do; that is, what is your job called? For example, electrical engineer, construction worker, carpenter, high school teacher, typist, etc.
- ○ I had no civilian job in 1991, GO TO QUESTION 122

WRITE THE NAME OF YOUR JOB IN THE BOX BELOW.

KIND OF WORK/JOB TITLE:

110. Which of the following best describes your civilian employer in 1991? Mark one.
- ○ Federal Government
- ○ State Government
- ○ Local Government (including public schools)
- ○ Self-employed in own business
- ○ Private firm with more than 500 employees
- ○ Private firm with 100-499 employees
- ○ Private firm with less than 100 employees
- ○ Working without pay in family business or farm

111. What kind of organization did you work for in 1991? (For example, TV and radio, manufacturing, retail shoe store, police department, etc. Federal workers: enter the Agency, Department or Government Branch for which you work.)

WRITE THE KIND OF ORGANIZATION (BUSINESS/INDUSTRY) IN THE BOX BELOW. DO NOT WRITE THE NAME OF THE COMPANY.

KIND OF ORGANIZATION:

112. What was your Federal Government pay type and grade at the end of 1991? Mark both the pay type and number grade.
- ○ Does not apply, I didn't work for the Federal Government

A. Pay Type	B. Number Grade	
○ SES or other executive pay	○ 16 or higher	○ 8
○ GM	○ 15	○ 7
○ GS	○ 14	○ 6
○ WS	○ 13	○ 5
○ WL	○ 12	○ 4
○ WG	○ 11	○ 3
○ US Postal Service	○ 10	○ 2
○ Other	○ 9	○ 1

USE NO. 2 PENCIL ONLY

113. **In 1991, how many hours per week did you usually work at your (main) civilian job?**

Hours Per Week Usually Worked

```
0 0
1 1
2 2
3 3
4 4
5 5
6 6
7 7
8 8
9 9
```

114. **In 1991, how often did you work more than 40 hours per week at your (main) civilian job?** Give your best estimate.
- ○ None
- ○ 1-4 weeks
- ○ 5-9 weeks
- ○ 10-14 weeks
- ○ 15-19 weeks
- ○ 20 or more weeks

115. **In 1991, how were you paid when you worked over 40 hours a week?** Mark one.
- ○ Not paid extra for working over 40 hours
- ○ Paid at my regular pay rate for all hours I worked
- ○ Paid time-and-a-half
- ○ Paid double time
- ○ Paid more than double time

116. **In 1991, what were your USUAL WEEKLY EARNINGS from your (main) civilian job or your own business before taxes and other deductions?** Give your best estimate.

Weekly Earnings

```
$ |  |  |  |  | .00
  0 0 0 0
  1 1 1 1
  2 2 2 2
  3 3 3 3
  4 4 4 4
  5 5 5 5
  6 6 6 6
  7 7 7 7
  8 8 8 8
  9 9 9 9
```

117. **In 1991, how many days of paid vacation did you receive from your (main) civilian job?**

○ I didn't receive paid vacation

Days of Paid Vacation

```
0 0
1 1
2 2
3 3
4 4
5 5
6 6
7
8
9
```

118. **In 1991, did you lose opportunities for overtime/extra pay because of your Guard/Reserve obligations?**
- ○ Yes, frequently
- ○ Yes, occasionally
- ○ No

119. **Which of the following describes how you got time off from your civilian job to meet your Guard/Reserve obligations in 1991?** Mark all that apply in each column.

○ Does not apply, I was self-employed, GO TO QUESTION 121

	OBLIGATIONS		
	A. Required Drills	B. Annual Training/ ACDUTRA	C. Military Schooling
Does not apply, I did not attend	○	○	○
I received military leave/leave of absence	○	○	○
I used vacation days	○	○	○
My Guard/Reserve obligations were on days on which I did not work	○	○	○

120. **Which of the following describes how you were paid for the time you took from your civilian job for Guard/Reserve obligations in 1991?** Mark all that apply in each column.

	OBLIGATIONS	
	A. Required Drills	B. Annual Training/ ACDUTRA
Does not apply, I did not attend	○	○
I received full civilian pay as well as military pay	○	○
I received partial civilian pay as well as military pay	○	○
I received only military pay	○	○
My Guard/Reserve obligations were on days on which I did not work	○	○

USE NO. 2 PENCIL ONLY

121. **During 1991, what was the TOTAL AMOUNT THAT YOU EARNED FROM ALL CIVILIAN JOBS or your own business BEFORE taxes and other deductions?** Include earnings as a Guard/Reserve technician. Include commissions, tips, or bonuses. Give your best estimate.

Amount Earned at Civilian Job

○ More than $100,000
○ None

$ ⬚⬚⬚⬚⬚ .00

(0)(0)(0)(0)(0)
(1)(1)(1)(1)(1)
(2)(2)(2)(2)(2)
(3)(3)(3)(3)(3)
(4)(4)(4)(4)(4)
(5)(5)(5)(5)(5)
(6)(6)(6)(6)(6)
(7)(7)(7)(7)(7)
(8)(8)(8)(8)(8)
(9)(9)(9)(9)(9)

122. **In 1991, how many weeks were you without a job and looking for work?**

Weeks Looking for Work

○ I had a job throughout 1991
○ I was not looking for work

(0)(0)
(1)(1)
(2)(2)
(3)(3)
(4)(4)
(5)(5)
(6)
(7)
(8)
(9)

123. **Do you currently have a spouse?**
○ No, GO TO QUESTION 131
○ Yes
○ Yes, separated, GO TO QUESTION 131

B. YOUR SPOUSE'S WORK EXPERIENCE

124. **Is your spouse: Mark all that apply.**
○ Working full-time in Federal civilian job
○ Working full-time in civilian job (not technician or Federal)
○ Working part-time in Federal civilian job
○ Working part-time in civilian job (not Federal)
○ Self-employed in his or her own business
○ With a job, but not at work because of TEMPORARY illness, vacation, strike, etc.
○ Unpaid worker (volunteer or in family business)
○ Unemployed, laid off, or looking for work
○ In school
○ Retired
○ A homemaker
○ Other

125. **Is your spouse: Mark all that apply.**
○ In the Armed Forces, full-time Active Component, GO TO QUESTION 126
○ In the Armed Forces, full-time Reserve Component (FTS-AGR/TAR), GO TO QUESTION 126
○ Full-time as a Guard/Reserve technician in the Army or the Air Force, GO TO QUESTION 127
○ Part-time in the Guard/Reserve, GO TO QUESTION 127
○ None of the above, GO TO QUESTION 129

126. **Was your full-time active duty spouse deployed during Operation Desert Shield/Desert Storm?**
○ No, re·ained at home installation, GO TO QUESTION 129
○ Yes, deployed to the Persian Gulf Area, GO TO QUESTION 128
○ Yes, deployed to other overseas location, GO TO QUESTION 128

127. **Was your Guard/Reserve spouse mobilized/ activated/called-up for Operation Desert Shield/Desert Storm?**
○ No, GO TO QUESTION 129
○ Yes, deployed to the Persian Gulf area
○ Yes, deployed to other overseas location
○ Yes, stayed in our local community
○ Yes, served elsewhere in United States

128. **How many months was your spouse on Active Duty during Operation Desert Shield/Desert Storm?**

Months

(0)(0)
(1)(1)
(2)(2)
(3)(3)
(4)(4)
(5)(5)
(6)(6)
(7)(7)
(8)(8)
(9)(9)

USE NO. 2 PENCIL ONLY

129. In 1991, how many hours per week did YOUR SPOUSE work for pay, either full or part-time, at a civilian job? Give your best estimate.

Hours Per Week

○ None, GO TO QUESTION 131

⓪⓪
①①
②②
③③
④④
⑤⑤
⑥⑥
⑦⑦
⑧⑧
⑨⑨

130. Altogether in 1991, what was the total amount that YOUR SPOUSE earned from a civilian job or his or her own business, **BEFORE taxes and other deductions**? Include earnings as a Guard/Reserve technician. Include commissions, tips, or bonuses. Give your best estimate.

Amount Earned by Spouse

○ More than $100,000
○ None

$ _____ .00

⓪⓪⓪⓪⓪
①①①①①
②②②②②
③③③③③
④④④④④
⑤⑤⑤⑤⑤
⑥⑥⑥⑥⑥
⑦⑦⑦⑦⑦
⑧⑧⑧⑧⑧
⑨⑨⑨⑨⑨

VII FAMILY RESOURCES

131. During 1991, did you or your spouse receive any income from the following sources? Mark "YES" or "NO" for each item.

RECEIVED

Yes	No	INCOME SOURCE
○	○	a. Interest and Dividends on Savings
○	○	b. Stocks, Bonds or Other Investments
○	○	c. Alimony, Child Support or Other Regular Contributions from Persons not Living in Your Household
○	○	d. Unemployment Compensation or Workers Compensation
○	○	e. Pensions from Federal, State or Local Government Employment
○	○	f. Pensions from Private Employer or Union
○	○	g. Veterans benefits or pensions
○	○	h. GI Bill
○	○	i. Social Security or Railroad Retirement
○	○	j. Supplemental Security Income
○	○	k. Public Welfare or Assistance
○	○	l. WIC (food programs for women, infants and children)
○	○	m. Government Food Stamps
○	○	n. Anything else not including earnings from wages or salaries

132. During 1991, how much did you or your spouse receive from the income sources listed in Question 131? Do not include earnings from wages or salaries in this question. Give your best estimate.

○ No income from sources in Question 131

$ _____ .00

○ More than $100,000

⓪⓪⓪⓪⓪
①①①①①
②②②②②
③③③③③
④④④④④
⑤⑤⑤⑤⑤
⑥⑥⑥⑥⑥
⑦⑦⑦⑦⑦
⑧⑧⑧⑧⑧
⑨⑨⑨⑨⑨

133. Overall how do you feel about your/your family income; that is, all the money that comes to you and other members of your family living with you?

○ Very satisfied
○ Satisfied
○ Neither satisfied nor dissatisfied
○ Dissatisfied
○ Very dissatisfied

USE NO. 2 PENCIL ONLY

YOUR RESIDENCE

134. How far is your new principal residence from your last principal residence? Mark one.
- ○ I have not moved since joining the Guard/Reserve
- ○ Less than 50 miles
- ○ 50 to 100 miles
- ○ 101 to 250 miles
- ○ 251 to 500 miles
- ○ More than 500 miles

135. Do you RENT or OWN your principal residence?
- ○ Neither, live in government-owned or leased housing
- ○ Neither, live with friends/relatives and PAY NO COSTS, GO TO QUESTION 142
- ○ Neither, live in other accommodations
- ○ RENT
- ○ OWN

136. How long have you RENTED or OWNED your residence?

○ 3 months or less	○ 37 to 48 months
○ 4 to 6 months	○ 49 to 59 months
○ 7 to 12 months	○ 5 to 10 years
○ 13 to 24 months	○ 11 to 20 years
○ 25 to 36 months	○ 21 or more years

If "RENT" continue with Question 137
If "OWN" go to Question 138

137. How much TOTAL RENT is paid for your residence PER MONTH?

If you share the rent, enter the total rent paid by all occupants. (For example, if it is $525 enter 0525 in the boxes and fill in the matching circles. Include RENT only. Other housing costs will be asked for later.)

Dollars Per Month

$ [ ][ ][ ][ ] .00

⓪⓪⓪⓪
①①①①
②②②②
③③③③
④④④
⑤⑤⑤
⑥⑥⑥
⑦⑦⑦
⑧⑧⑧
⑨⑨⑨

138. What is your monthly house payment for your residence? (Include the PRINCIPAL AND INTEREST on all mortgages or trusts, real estate TAXES and homeowner's INSURANCE. Also include land lease, mobile home lot rental, or berthing fees, if applicable. Other housing costs, such as utility and maintenance costs, etc., will be asked for later. Example: if your payment is $890, enter 0890 in the boxes, then fill in the matching circles.)

Dollars Per Month

$ [ ][ ][ ][ ] .00

⓪⓪⓪⓪
①①①①
②②②②
③③③③
④④④
⑤⑤⑤
⑥⑥⑥
⑦⑦⑦
⑧⑧⑧
⑨⑨⑨

139. Over the last 12 months, what was the AVERAGE MONTHLY cost of all utilities (except telephone and cable TV) paid separately from other rental or home ownership costs?
- ○ DOES NOT APPLY, No utilities are paid separately
- ○ Do not have a basis for estimating utility costs

For each utility, add all costs for the LAST 12 MONTHS and divide by 12. (If you do not know the costs for all 12 months, please estimate.)

Enter the average monthly cost for each utility in the space below, then enter the TOTAL at the right.

Dollars Per Month

→ $ [ ][ ][ ][ ] .00

	Monthly Average
Electricity	$
Natural Gas/Propane	$
Fuel Oil	$
Wood/Coal	$
Water/Sewer	$
Garbage	$
Total	$

⓪⓪⓪
①①①
②②②
③③③
④④④
⑤⑤⑤
⑥⑥
⑦⑦
⑧⑧
⑨⑨

USE NO. 2 PENCIL ONLY

140. Enter the AVERAGE MONTHLY maintenance cost paid for the UPKEEP of the residence. Round off to the nearest dollar.

○ No maintenance costs are paid separately

Dollars Per Month

$ [ ][ ][ ] .00

⓪⓪⓪
①①①
②②②
③③③
④④④
⑤⑤
⑥⑥
⑦⑦
⑧⑧
⑨⑨

- INCLUDE only maintenance such as plumbing, electrical, heating/cooling system or structural repairs, yard upkeep, etc.
- DO NOT INCLUDE the cost of home improvements (e.g., remodeling, new roof, new furnace, major appliances), new shrubs, new fences, or other additions.
  Example: If your cost is $25 per month, enter 025 in the boxes, then fill in the matching circles.

141. Enter the AVERAGE MONTHLY cost of any of the following housing expenses for the residence, condominium fee, homeowner's association fee, property and hazard insurance, if NOT included in Question 137 or Question 138.

Fill in the grid for EACH expense you do have or mark "None" for EACH expense you do not have.

	Condominium Fee	Homeowner's Assoc. Fee	Property & Hazard Insurance
	○ None	○ None	○ None
Dollars per Month	$ [ ][ ][ ]	$ [ ][ ][ ]	$ [ ][ ][ ]

Write the numbers in the boxes

Then fill in the matching circles

⓪⓪⓪ ⓪⓪⓪ ⓪⓪⓪
①①① ①①① ①①①
②②② ②②② ②②②
③③③ ③③③ ③③③
④④④ ④④④ ④④④
⑤⑤⑤ ⑤⑤⑤ ⑤⑤⑤
⑥⑥ ⑥⑥ ⑥⑥⑥
⑦⑦ ⑦⑦ ⑦⑦⑦
⑧⑧ ⑧⑧ ⑧⑧⑧
⑨⑨ ⑨⑨ ⑨⑨⑨

VIII MILITARY LIFE

142. How do you feel about the amount of time you spend on each activity listed below? Mark one for each activity.

	I Spend Too Much Time	I Spend About the Right Amount of Time	I Don't Spend Enough Time	Does Not Apply
a. Your civilian job	○	○	○	○
b. Family activities	○	○	○	○
c. Leisure activities	○	○	○	○
d. Guard/Reserve activities	○	○	○	○
e. Community activities	○	○	○	○

143. The Guard/Reserve are developing new information materials. Below is a list of topics that might be included. How interested would you be in receiving such materials? Please mark your interest in information about each topic.

For each item, mark if you are:	Very Interested	Interested	Somewhat Interested	Not Interested At All
a. Retirement benefits	○	○	○	○
b. Survivor Benefit Plan	○	○	○	○
c. Family benefits in the Guard/Reserve	○	○	○	○
d. Mobilization procedures for dependents	○	○	○	○
e. Selected Reserve GI Bill Educational Assistance	○	○	○	○
f. Soldiers/Sailors Civil Relief	○	○	○	○
g. Dental Insurance	○	○	○	○
h. Medical Insurance	○	○	○	○
i. Mobilization Preparations for Small Business Owners and Partners/Independent Practitioners	○	○	○	○

USE NO. 2 PENCIL ONLY

144. All things considered, please indicate your level of satisfaction or dissatisfaction with each feature of the Guard/Reserve listed below.

For each item, mark if you are:

	Very Satisfied	Satisfied	Neither Satisfied Nor Dissatisfied	Dissatisfied	Very Dissatisfied
a. Military pay and allowances	○	○	○	○	○
b. Commissary privileges	○	○	○	○	○
c. Exchange privileges	○	○	○	○	○
d. Morale/welfare/recreation privileges	○	○	○	○	○
e. Time required at Guard/Reserve activities	○	○	○	○	○
f. Military retirement benefits	○	○	○	○	○
g. Unit social activities	○	○	○	○	○
h. Opportunities for education/training	○	○	○	○	○
i. Opportunity to serve one's country	○	○	○	○	○
j. Acquaintances/friendships	○	○	○	○	○

145. Overall, how satisfied are you with the pay and benefits you receive for the amount of time you spend on Guard/Reserve activities?

Very Dissatisfied   Very Satisfied

①—②—③—④—⑤—⑥—⑦

146. Overall, how satisfied are you with your participation in the Guard/Reserve?

Very Dissatisfied   Very Satisfied

①—②—③—④—⑤—⑥—⑦

147. We're interested in any comments you'd like to make about Guard/Reserve personnel policies, whether or not the topic was covered in this survey.

DO YOU HAVE ANY COMMENTS?

○ No
○ Yes – Please fill out the COMMENT SHEET on page 23.

THANK YOU VERY MUCH FOR ANSWERING THIS SURVEY. PLEASE RETURN IT IN THE ENVELOPE PROVIDED.

COMMENT SHEET

Please provide us with comments you may have regarding Reserve policies or Reserve activities in general in the space below. Before commenting, please fill in one circle in each section.

Your Rank

○ Officer
○ Enlisted

Your Component

○ Army National Guard (ARNG)
○ Army Reserve (USAR)
○ Naval Reserve (USNR)
○ Marine Corps Reserve (USMCR)
○ Air National Guard (ANG)
○ Air Force Reserve (USAFR)
○ Coast Guard Reserve (USCGR)

1992 RESERVE COMPONENTS SURVEY OF OFFICERS

RCS DD—FM & P (OT) 1952

1992 Reserve Components
Survey of Officers

The National Guard and Reserve Components are conducting a survey of Guard/Reserve personnel. You have been selected to participate in this important survey. Please read the instructions before you begin the questionnaire.

PRIVACY NOTICE

AUTHORITY: 10 U.S.C. 136

PRINCIPAL PURPOSE OR PURPOSES:
Information collected in this survey is used to sample attitudes and/or discern perceptions of social problems observed by the Guard and Reserve Components members and to support additional manpower research activities. This information will assist in the formulation of policies which may be needed to improve the environment for Reserve Components members and families.

ROUTINE USES: None

DISCLOSURE: Your survey instrument will be treated as confidential. All identifiable information will be used only by persons engaged in, and for the purposes of, the survey. It will not be disclosed to others or used for any other purpose. Only group statistics will be reported.

Your participation in the survey is voluntary. Failure to respond to any questions will not result in any penalty. However, your participation is encouraged so that the data will be complete and representative.

```
OFFICE USE ONLY
O PN
O NR
O RF
O NE
```

INSTRUCTIONS FOR COMPLETING THE SURVEY

use use a No. 2 pencil.

Make heavy black marks that fill the circle for your answer.
- Please do not make stray marks of any kind.

INCORRECT MARKS   CORRECT MARK

○ 🅇 ℯ ○   ○ ● ○ ○

- Sometimes you will be asked to "Mark one." When this instruction appears, mark the one best answer.

Example:
In what month are you completing the survey?
- ○ August
- ● September
- ○ October
- ○ November
- ○ December
- ○ January
- ○ February

your answer is "September," then just mark that one circle.

- Sometimes you will be asked to "Mark all that apply." When this instruction appears, you may mark more than one answer.

Example:
In which components have you served? Mark all that apply.
- ● Active Army (USA)
- ◐ Army National Guard (ARNG)
- ● Army Reserve (USAR)
- ○ Active Navy (USN)
- ○ Naval Reserve (USNR)
- ○ Active Air Force (USAF)
- ○ Air National Guard (ANG)
- ○ Air Force Reserve (USAFR)
- ○ Active Marine Corps (USMC)
- ○ Marine Corps Reserve (USMCR)
- ○ Active Coast Guard (USCG)
- ○ Coast Guard Reserve (USCGR)

If your answer is "Active Army (USA)" and "Army serve (USAR)," then mark the two circles clearly.

- Answers to some of the questions will be on a SEVEN-POINT SCALE.

Example:
How satisfied are you with the opportunities you have for promotion in your unit?

Very
Dissatisfied   Very
                                          Satisfied

①—②—③—●—⑤—⑥—⑦

If your answer is "VERY DISSATISFIED," you would darken the circle for number 1.

If your answer is "VERY SATISFIED," you would darken the circle for number 7.

If your opinion is somewhere in between, you would darken the circle for number 2 or 3 or 4 or 5 or 6.

- If you are asked to give numbers for your answer, please record as shown below.

Example:
How old were you on your last birthday?

If your answer is 24...
Write the numbers in the boxes, making sure that the last number is always placed in the right-hand box.

Fill in the unused boxes with zeros.

Then darken the circle for the matching number below each box.

Age Last Birthday

Write the number in the boxes. → [2 | 4]

                    ⓪ ⓪
                    ① ①
Then fill in the → ② ②
matching circles.   ③ ③
                    ④ ●
                    ⑤ ⑤
                    ⑥ ⑥
                    ⑦ ⑦
                    ⑧ ⑧
                    ⑨ ⑨

USE NO. 2 PENCIL ONLY

I LOCATION

In what month are you completing the survey?
Mark one.

- ○ August
- ○ September
- ○ October
- ○ November
- ○ December
- ○ January
- ○ February

2. Which of the following best describes the type of place where you are living now? Mark one.

- ○ In military housing on a base/installation
- ○ In a large city (over 250,000)
- ○ In a suburb near a large city
- ○ In a medium-sized city (50,000-250,000)
- ○ In a suburb near a medium-sized city
- ○ In a small city or town (under 50,000)
- ○ On a farm or ranch
- ○ In a rural area but not on a farm or ranch

3. How long have you lived in your present neighborhood? Mark one.

- ○ Less than a year
- ○ 1-2 years
- ○ 2-3 years
- ○ 3-5 years
- ○ 5 years or more

II MILITARY BACKGROUND

4. Of which Reserve Component are you a member? Mark one.

- ○ Army National Guard (ARNG)
- ○ Army Reserve (USAR)
- ○ Naval Reserve (USNR)
- ○ Marine Corps Reserve (USMCR)
- ○ Air National Guard (ANG)
- ○ Air Force Reserve (USAFR)
- ○ Coast Guard Reserve (USCGR)

5. What is your present pay grade? Mark one.

○ O-1 or O1-E	○ O-5	W-1
○ O-2 or O2-E	○ O-6	W-2
○ O-3 or O3-E	○ O-7 or above	W-3
○ O-4		W-4

6. When do you expect to get your NEXT PROMOTION to a higher pay grade? Mark one.

- ○ In less than 3 months
- ○ 3-6 months from now
- ○ 7-9 months from now
- ○ 10-12 months from now
- ○ 13-18 months from now
- ○ 19 months to 2 years from now
- ○ 25 months to 3 years from now
- ○ More than 3 years from now
- ○ Does not apply, I don't expect any more promotions

7. In what year did you first enter any branch of the military? (If you first entered in the Active Force, record the year you first entered the Active Force.)

Write the number in the boxes. → 19 ☐ ☐

Then fill in the → matching circles.

⓪ ⓪
① ①
② ②
③ ③
④ ④
⑤ ⑤
⑥ ⑥
⑦ ⑦
⑧ ⑧
⑨ ⑨

8. Through which of the following officer procurement programs did you obtain your commission/warrant? Mark one.

- ○ Academy Graduate (USMA, USNA, USAFA, USCGA)
- ○ Academy Graduate (U.S. Merchant Marine Academy)
- ○ ROTC/NROTC (scholarship)
- ○ ROTC/NROTC (non-scholarship)
- ○ OCS/AOCS/OTS/PLC
- ○ Aviation Cadet
- ○ National Guard State OCS
- ○ ANG Academy of Military Science (AMS)
- ○ Direct appointment (professional-medical, dental, JAG, chaplain)
- ○ Direct appointment (all others)
- ○ Aviation training program (exclusive of OCS/AOCS/OTS/PLC)
- ○ Direct appointment as a commissioned officer
- ○ Direct appointment as a warrant officer
- ○ Warrant Officer Entry Level Training
- ○ Other

USE NO. 2 PENCIL ONLY

9. When you first entered the military, **in which** component did you serve? Do not include as active service, service for basic and initial training only. Mark one.

- ○ Active Army (USA)
- ○ Army National Guard (ARNG)
- ○ Army Reserve (USAR)
- ○ Active Navy (USN)
- ○ Naval Reserve (USNR)
- ○ Active Air Force (USAF)
- ○ Air National Guard (ANG)
- ○ Air Force Reserve (USAFR)
- ○ Active Marine Corps (USMC)
- ○ Marine Corps Reserve (USMCR)
- ○ Active Coast Guard (USCG)
- ○ Coast Guard Reserve (USCGR)

10. In which components have you served? Do not include as active service, service for basic and initial training only. Mark all that apply.

- ○ Active Army (USA)
- ○ Army National Guard (ARNG)
- ○ Army Reserve (USAR)
- ○ Active Navy (USN)
- ○ Naval Reserve (USNR)
- ○ Active Air Force (USAF)
- ○ Air National Guard (ANG)
- ○ Air Force Reserve (USAFR)
- ○ Active Marine Corps (USMC)
- ○ Marine Corps Reserve (USMCR)
- ○ Active Coast Guard (USCG)
- ○ Coast Guard Reserve (USCGR)

11. In all, to the nearest year, how long have you served in the Guard/Reserve? Do not include active duty years.

○ Less than 1 year

12. In all, to the nearest year, how long did you serve in the Active Force/ on active duty? Do not include your initial active duty training for the Guard/Reserve. Include service as FTS-AGR/TAR.

- ○ I have never served in the Active Force
- ○ Less than 1 year

13. When you finally leave the Guard/Reserve, how many total years of service do you expect to have? (Include active duty years.)

14. Are you in a different unit now than you were two years ago? Mark one.

- ○ I have not been in the Guard/Reserve for two years, GO TO QUESTION 17
- ○ No, I am in the same unit, GO TO QUESTION 17
- ○ Yes, in a different unit but in the same component
- ○ Yes, in a different unit in a different component

15. Why did you change units? Mark all that apply.

- ○ I was offered a promotion
- ○ Promotion was more likely in new unit
- ○ I relocated away from the previous unit
- ○ I wanted to retrain in a different skill
- ○ I like the job better in my new unit
- ○ I like the people better in my new unit
- ○ My old unit was disestablished
- ○ Other reasons

USE NO. 2 PENCIL ONLY

16. Did you have to retrain in a new skill when you changed units?
- O Yes
- O No

17. Were you mobilized/activated/called-up as a Reservist during Operation Desert Shield/Desert Storm?
Mark all that apply.
- O No, GO TO QUESTION 19
- O Yes, deployed to Persian Gulf area
- O Yes, deployed to other overseas location
- O Yes, deployed in the United States
- O Yes, stayed in my local community

18. How many months were you mobilized/activated/called-up?

Number Months

0	0
1	1
2	2
3	3
4	4
	5
	6
	7
	8
	9

III MILITARY PLANS

19. When you originally became a member of the Guard/Reserve, how many years were you obligated to serve? Mark one.
- O No original obligation
- O 1 year or less
- O 2 years
- O 3 years
- O 4 years
- O 5 years
- O 6 years
- O 7 years
- O 8 years
- O Don't know

20. Do you have a current obligation or term of service?
- O Yes
- O No, GO TO QUESTION 23

---

21. In what month and year will you complete your current obligation or term of service?

A	B
Month	Year
O January	199
O February	
O March	0
O April	1
O May	2
O June	3
O July	4
O August	5
O September	6
O October	7
O November	8
O December	9
O Don't know	

22. At the completion of your obligation or term of service, how likely are you to continue to participate in the Selected Reserve of the Guard/Reserve?
- O (0 in 10) No chance
- O (1 in 10) Very slight possibility
- O (2 in 10) Slight possibility
- O (3 in 10) Some possibility
- O (4 in 10) Fair possibility
- O (5 in 10) Fairly good possibility
- O (6 in 10) Good possibility
- O (7 in 10) Probable
- O (8 in 10) Very probable
- O (9 in 10) Almost sure
- O (10 in 10) Certain

23. When you finally leave the Guard/Reserve, what pay grade do you think you will have? Mark one.
- O O-1
- O O-2
- O O-3
- O O-4
- O O-5
- O O-6
- O O-7 or above
- O W-1
- O W-2
- O W-3
- O W-4
- O W-5

USE NO. 2 PENCIL ONLY

24. Below are some reasons people have for DECIDING TO LEAVE the National Guard/Reserve. If you decide to leave the Guard/Reserve at the end of your current term, which of these would be your most important reason for leaving? Which would be your second most important reason for leaving? (Mark one reason under each column.)

I WOULD LEAVE THE GUARD/RESERVE BECAUSE:	(A) Most Important Reason	(B) Second Most Important Reason
a. I am not eligible to reenlist	○	○
b. I am moving to another area	○	○
c. It is too hard to get to my Guard/Reserve unit	○	○
d. I need the time for my education	○	○
e. My unit drills conflict with my civilian job	○	○
f. My unit drills conflict with my family activities	○	○
g. I want more leisure time	○	○
h. I don't like my unit's training	○	○
i. My unit doesn't have modern equipment for training	○	○
j. I'm bored with unit activities	○	○
k. The pay is too low	○	○
L. Promotions are too slow	○	○
m. I've had too many problems getting paid	○	○
n. Problems caused by mobilization/activation/deployment	○	○

25. How likely are you to stay in the Guard/Reserve until qualified for retirement? Assume that all special pays which you currently receive are still available. Mark one.

- ○ (0 in 10) No chance
- ○ (1 in 10) Very slight possibility
- ○ (2 in 10) Slight possibility
- ○ (3 in 10) Some possibility
- ○ (4 in 10) Fair possibility
- ○ (5 in 10) Fairly good possibility
- ○ (6 in 10) Good possibility
- ○ (7 in 10) Probable
- ○ (8 in 10) Very probable
- ○ (9 in 10) Almost sure
- ○ (10 in 10) Certain

26. Do you plan to elect the Reserve Components Survivor Benefit Plan (SBP) when eligible?

- ○ Does not apply, I don't plan to remain until 20 years
- ○ I have already elected to participate
- ○ I have already elected not to participate
- ○ Yes, upon receipt of my 20-year letter
- ○ Yes, when I am 60 years old
- ○ No
- ○ Uncertain, I am not aware of the plan at all
- ○ Uncertain, I don't understand the plan clearly
- ○ Uncertain, I have not made up my mind

27. How concerned are you about the following as a result of current talk about force reductions in the Guard/Reserve? Mark one for each item.

	Very Greatly Concerned	Greatly Concerned	Moderately Concerned	Somewhat Concerned	Not At All Concerned
a. Your long-term opportunities in the Guard/Reserve	○	○	○	○	○
b. The financial burden on you and/or your family should you have to leave the Guard/Reserve unexpectedly	○	○	○	○	○
c. Impact of my unit closing on my community	○	○	○	○	○

28. The questions below are about your preparedness. Mark one for each item.

	Yes	No	Don't Know	Does Not Apply
a. Do you have a current written will?	○	○	○	○
b. Does anyone currently hold your power-of-attorney?	○	○	○	○
c. Do you have life insurance other than SGLI/VGLI?	○	○	○	○
d. Have you filled out a record of emergency data?	○	○	○	○
e. Does your spouse or next-of-kin know where to find your papers?	○	○	○	○
f. Do you verify/update annually your record of emergency data?	○	○	○	○

USE NO. 2 PENCIL ONLY

29. If you were to be called up, how much of a problem would each of the following be for you or your family? Mark one number for each item.

	A Serious Problem						Not A Problem	Don't Know	Does Not Apply
¹¹ a. Employer problems at the beginning of the mobilization/activation/call-up	①	②	③	④	⑤	⑥	⑦	○	○
b. Employer problems when you returned to your job	①	②	③	④	⑤	⑥	⑦	○	○
c. Getting the same job back after returning	①	②	③	④	⑤	⑥	⑦	○	○
d. Loss of civilian health benefits during the call-up	①	②	③	④	⑤	⑥	⑦	○	○
e. Loss of seniority, promotion opportunity, or job responsibility on civilian job	①	②	③	④	⑤	⑥	⑦	○	○
f. Loss of income during the call-up	①	②	③	④	⑤	⑥	⑦	○	○
g. Attitudes of supervisor or co-workers upon return	①	②	③	④	⑤	⑥	⑦	○	○
h. Business or medical practice would be damaged	①	②	③	④	⑤	⑥	⑦	○	○
i. Problems for patients, clients, customers	①	②	③	④	⑤	⑥	⑦	○	○
j. Spouse would need work but would not find job	①	②	③	④	⑤	⑥	⑦	○	○
k. Increased family problems	①	②	③	④	⑤	⑥	⑦	○	○
l. Increased chances for a marital separation or divorce	①	②	③	④	⑤	⑥	⑦	○	○
m. Problems for children	①	②	③	④	⑤	⑥	⑦	○	○
n. Burden on spouse	①	②	③	④	⑤	⑥	⑦	○	○
o. Child care during the call-up	①	②	③	④	⑤	⑥	⑦	○	○

30. People participate in the Guard/Reserve for many reasons. How much have each of the following contributed to your most recent decision to stay in the Guard/Reserve? Mark one for each item.

	Major Contribution	Moderate Contribution	Minor Contribution	No Contribution
a. Serving the country	○	○	○	○
b. Using educational benefits	○	○	○	○
c. Obtaining training in a skill that would help get a civilian job	○	○	○	○
d. Serving with the people in the unit	○	○	○	○
e. Getting credit toward Guard/Reserve retirement	○	○	○	○
f. Promotion opportunities	○	○	○	○
g. Opportunity to use military equipment	○	○	○	○
h. Challenge of military training	○	○	○	○
i. Needed the money for basic family expenses	○	○	○	○
j. Wanted extra money to use now	○	○	○	○
k. Saving income for the future	○	○	○	○
l. Travel "get away" opportunities	○	○	○	○
m. Just enjoyed the Guard/Reserve	○	○	○	○
n. Pride in my accomplishments in the Guard/Reserve	○	○	○	○

USE NO. 2 PENCIL ONLY

IV MILITARY TRAINING, BENEFITS, AND PROGRAMS

How were you trained for your current Primary Occupational Specialty (MOS/Designator/Rating/AFSC)? Mark all that apply.

○ In a formal service school
○ On-the-job training (OJT) in a civilian job
○ In a formal civilian school
○ On-the-job training (OJT) in the active service
○ On-the-job training (OJT) in a Guard/Reserve unit
○ Correspondence course(s)

2. For all of 1991, what percentage of your Guard/Reserve time was spent working in your Primary Occupational Specialty (MOS/Designator/Rating/AFSC)?

○ None   ○ 25-49%   ○ 75-99%
○ 1-24%   ○ 50-74%   ○ 100% (All)

3. Is your current Primary Occupational Specialty (MOS/Designator/Rating/AFSC) the same one you had while on active duty?

○ Does not apply. I don't have   ○ Yes
    prior active duty service   ○ No

4. How similar is your civilian job to your Guard/Reserve duty?

○ Does not apply. I don't have a civilian job
○ Does not apply, my civilian job is as a Guard/Reserve military technician
○ Very similar
○ Similar
○ Somewhat similar
○ Not similar at all

35. In calendar year 1991, which of the following did you participate in/perform? Mark all that apply.

○ Drill weekends
○ Annual Training/ACDUTRA
○ Active duty (other than for training)
○ Active duty for school training
○ Guard/Reserve work at my home or on my civilian job

36. In 1991, how many days of Annual Training/ACDUTRA did you attend? Do not include school unless used to satisfy your Annual Training/ACDUTRA requirement.

○ Did not attend 1991 Annual Training/ACDUTRA

Days

37. Did you attend the 1991 Annual Training/ACDUTRA a few days at a time, a week or more at a time, or all at once?

○ Did not attend 1991 Annual Training/ACDUTRA
○ A few days at a time, several times over the year
○ A week or more at a time
○ All at once

38. In calendar year 1991, how many paid "Workdays," in addition to any regular drill days and Annual Training/ACDUTRA, did you serve?

○ None

Paid Workdays

39. In an average month in 1991, how many unpaid hours did you spend at your drill location (place of regular duty)?

○ None

Unpaid Hours Per Month

40. For all of 1991, what was your total Guard/Reserve Income BEFORE taxes and deductions? Include any pay from drills, Annual Training/ACDUTRA, enlistment or affiliation bonuses, and any call-ups or other active duty or active duty for training. Please give your best estimate.

• Record the amount in the boxes. ➞ $ ☐☐☐☐☐ .00

• Round to the nearest whole dollar.

• Fill in the unused boxes with zeros. (For example, if your answer is $1,503.75, enter 01504.)

• Then mark the matching circle below each box. ➞

Total Guard/Reserve Income

USE NO. 2 PENCIL ONLY

41. In an average month in 1991, how often did you and/or your spouse use each of the following? Mark one for each item.

TIMES USED IN AVERAGE MONTH

	Not Used	Once	Twice	Three to Five Times	Six Times or More
a. Commissary	O	O	O	O	O
b. Exchange	O	O	O	O	O
c. Other military facilities	O	O	O	O	O

42. Which of the following limit your and/or your spouse's use of the commissary and exchange? Mark all that apply in each column.

	A. Commissary	B. Exchange
Prices	O	O
Stock	O	O
Hours	O	O
Distance	O	O
Military does not allow more frequent use	O	O

43. Are you now eligible for educational benefits as a result of military service? Mark all that apply.
- O No, GO TO QUESTION 45
- O Yes, State benefits for my Guard/Reserve service
- O Yes, Montgomery GI Bill for Selected Reserve
- O Yes, Active Force benefits (VEAP, GI Bill)
- O Don't know/am not sure

44. Which educational benefits are you now using? Mark all that apply
- O None
- O State benefits for Guard/Reserve
- O Montgomery GI Bill for Selected Reserve
- O Active Force benefits (VEAP, GI Bill)

45. Which of the following medical/hospitalization coverages do you have? Mark all that apply
- O My spouse's active duty military coverage
- O My active duty military coverage
- O Veterans' (VA) coverage
- O My civilian employer's health care plan
- O My spouse's civilian employer's plan
- O Other private coverage
- O None, GO TO QUESTION 47

46. How would you rate the coverage provided by the civilian medical insurance which you have?
- O Does not apply, I do not have civilian medical insurance
- O Excellent
- O Good
- O Fair
- O Poor

47. If it were available through your membership in the Guard or Reserve, would you be interested in purchasing medical insurance?
- O Yes, for myself and my family
- O Yes, for myself only
- O Not sure
- O No, GO TO QUESTION 49

48. If you could buy medical insurance through Guard/Reserve participation, what is the maximum premium cost you would be willing to pay per month?
- O Less than $50 per month
- O $50 per month
- O $100 per month
- O $150 per month
- O $200 per month
- O $250 per month or more

49. How much did you spend on health care services and products (for you and your family) last year? Include CHAMPUS deductions, civilian insurance premiums, and drugs, etc. Do not include dental care.
- O Less than $100
- O $100 to $500
- O $501 to $1,000
- O $1,001 to $1,500
- O $1,501 to $2,500
- O More than $2,500
- O Don't know

50. Which of the following dental coverages do you have? Mark all that apply.
- O My spouse's active duty military coverage
- O My active duty military coverage
- O Veterans' (VA) coverage
- O My civilian employer's dental plan
- O My spouse's civilian employer's plan
- O Other private coverage
- O None, GO TO QUESTION 52

51. How would you rate the coverage provided by the civilian dental insurance which you have?
- O Does not apply, I do not have civilian dental insurance
- O Excellent
- O Good
- O Fair
- O Poor

52. If it were available through your membership in the Guard or Reserve, would you be interested in purchasing dental insurance?
- O Yes, for myself and my family
- O Yes, for myself only
- O Not sure
- O No, GO TO QUESTION 54

USE NO. 2 PENCIL ONLY

53. If you could buy dental insurance through monthly withholding from your Reserve paycheck, what is the maximum premium cost you would be willing to pay per month?

- ○ Less than $25 per month
- ○ $50 per month
- ○ $100 per month
- ○ $150 per month
- ○ $200 per month
- ○ $250 or more per month

54. How much did you spend for dental treatment (for you and your family) last year? (Include civilian premiums as well as direct payments for treatment.)

- ○ Less than $100
- ○ $100 – $200
- ○ $201 – $300
- ○ $301 – $500
- ○ $501 – $800
- ○ $801 – $1,000
- ○ $1,001 – $2,000
- ○ More than $2,000
- ○ Don't know

55. How much of a problem is each of the following for your unit in meeting your unit's training objectives? Please mark the number which shows your opinion on the lines below. For example, people who feel that an item is Not A Problem would mark 7. People who feel that an item is A Serious Problem would mark 1. Others may have opinions somewhere between 1 and 7. Mark one for each item.

	A Serious Problem						Not A Problem	Don't Know
a. Out-of-date equipment/weapons	①	②	③	④	⑤	⑥	⑦	○
b. Poor mechanical condition of equipment/weapons	①	②	③	④	⑤	⑥	⑦	○
c. Being below strength in Grades E-1 – E-4	①	②	③	④	⑤	⑥	⑦	○
d. Being below strength in Grades E-5 – E-9	①	②	③	④	⑤	⑥	⑦	○
e. Not enough staff resources to plan effective training	①	②	③	④	⑤	⑥	⑦	○
f. Low attendance of unit personnel at Unit Drills	①	②	③	④	⑤	⑥	⑦	○
g. Low attendance of unit personnel at Annual Training/ACDUTRA	①	②	③	④	⑤	⑥	⑦	○
h. Ineffective training during Annual Training/ACDUTRA	①	②	③	④	⑤	⑥	⑦	○
i. Shortage of MOS/Rating/Specialty/AFSC qualified personnel	①	②	③	④	⑤	⑥	⑦	○
j. Low quality of personnel in low grade unit drill positions	①	②	③	④	⑤	⑥	⑦	○
k. Not enough drill time to practice skills	①	②	③	④	⑤	⑥	⑦	○
l. Not enough time to plan training objectives and get all administrative paperwork done	①	②	③	④	⑤	⑥	⑦	○
m. Lack of access to good training facilities and grounds	①	②	③	④	⑤	⑥	⑦	○
n. Lack of good instruction manuals and materials	①	②	③	④	⑤	⑥	⑦	○
o. Lack of supplies, such as ammunition, gasoline, etc.	①	②	③	④	⑤	⑥	⑦	○
p. Excessive turnover of unit personnel	①	②	③	④	⑤	⑥	⑦	○
q. Inability to schedule effective unit annual training due to gaining command's operating schedule	①	②	③	④	⑤	⑥	⑦	○
r. Uncertainty about future status of unit	①	②	③	④	⑤	⑥	⑦	○

PLEASE CHECK. HAVE YOU MARKED A CIRCLE FOR EACH ITEM?

56. How do you usually get to the place of regular military duty or drills? Mark one.

- ○ Drive myself
- ○ Driven by spouse
- ○ Driven by another family member
- ○ Car pool
- ○ Civilian air transportation
- ○ Military air transportation
- ○ Other public transportation
- ○ Taxi
- ○ Walk
- ○ Other

57. How long does it usually take you to get from home to the place where your unit meets/drills? Mark one.

- ○ 0-19 minutes
- ○ 20-39 minutes
- ○ 40-59 minutes
- ○ 1-2 hours
- ○ 2-3 hours
- ○ 3-6 hours
- ○ 6 hours or more

FOR QUESTION 58 TO QUESTION 67 BELOW, PLEASE MARK THE NUMBER WHICH SHOWS YOUR OPINION ON THE LINE FOLLOWING EACH QUESTION. For example, people who are Very Satisfied would mark 7. People who are Very Dissatisfied would mark 1. Others may have opinions somewhere between 1 and 7.

58. How satisfied are you with the training received during your unit drills?

Very Dissatisfied ... Very Satisfied

①—②—③—④—⑤—⑥—⑦

59. How satisfied are you with the opportunities you have to use your MOS/Designator/Rating/Specialty/AFSC skills during unit drills?

Very Dissatisfied ... Very Satisfied

①—②—③—④—⑤—⑥—⑦

60. How satisfied are you with the opportunities you have for promotion in your unit?

Very Dissatisfied ... Very Satisfied

①—②—③—④—⑤—⑥—⑦

61. How satisfied are you with your opportunities for leadership in your unit?

Very Dissatisfied ... Very Satisfied

①—②—③—④—⑤—⑥—⑦

62. In general, how would you describe the weapons or equipment your unit uses during your unit drills?

Out-of-Date ... Up-to-Date

①—②—③—④—⑤—⑥—⑦

63. In general, how would you describe the mechanical condition of the weapons and equipment your unit uses during training?

Poor ... Excellent

①—②—③—④—⑤—⑥—⑦

64. Overall, how satisfied were you with your unit's activities at 1991 Annual Training/ACDUTRA?

○ Does not apply, I didn't attend 1991 Annual Training/ACDUTRA

Very Dissatisfied ... Very Satisfied

①—②—③—④—⑤—⑥—⑦

?. In general, how would you describe the morale of military personnel in your unit?

Morale is Very Low ... Morale is Very High

①—②—③—④—⑤—⑥—⑦

66. In general, how satisfied are you with the supervision and direction given during unit drills?

Very Dissatisfied ... Very Satisfied

①—②—③—④—⑤—⑥—⑦

67. How do you feel about not going to the Persian Gulf area during Operation Desert Storm/Desert Shield?

○ Does not apply, I went to the Persian Gulf area

Very Displeased ... Very Pleased

①—②—③—④—⑤—⑥—⑦

68. How long have you been in your present unit?

○ Less than 1 year

Years in Present Unit

⓪ ⓪
① ①
② ②
③ ③
④ ④
⑤
⑥
⑦
⑧
⑨

69. How likely is it that another conflict requiring a Reserve call-up will occur in the next 5 years?

○ (0 in 10) No chance
○ (1 in 10) Very slight possibility
○ (2 in 10) Slight possibility
○ (3 in 10) Some possibility
○ (4 in 10) Fair possibility
○ (5 in 10) Fairly good possibility
○ (6 in 10) Good possibility
○ (7 in 10) Probable
○ (8 in 10) Very probable
○ (9 in 10) Almost sure
○ (10 in 10) Certain

70. How likely is it that you would be called-up if such a mobilization occurred?

○ (0 in 10) No chance
○ (1 in 10) Very slight possibility
○ (2 in 10) Slight possibility
○ (3 in 10) Some possibility
○ (4 in 10) Fair possibility
○ (5 in 10) Fairly good possibility
○ (6 in 10) Good possibility
○ (7 in 10) Probable
○ (8 in 10) Very probable
○ (9 in 10) Almost sure
○ (10 in 10) Certain

USE NO. 2 PENCIL ONLY

1. If you were mobilized for 30 days or more, would your total income:
   ○ Increase greatly
   ○ Increase somewhat
   ○ Remain the same
   ○ Decrease somewhat
   ○ Decrease greatly

2. If mobilized, would you mobilize with your present unit?
   ○ Yes
   ○ No
   ○ Don't know

3. If mobilized, would your military duties be the same as your current duties when attending Annual Training/ACDUTRA?
   ○ Yes
   ○ No
   ○ Don't know

4. Are you Army or Air Force National Guard or Reserve?
   ○ Yes
   ○ No, GO TO QUESTION 78

5. Are you a military technician, i.e., a civilian employee of the Army or Air Force National Guard or Reserve?
   ○ Yes
   ○ No, GO TO QUESTION 78

6. How long have you been employed as a military technician?

   Years as Technician
   ○ Less than 1 year

77. Do you drill with the same unit that you work in as a technician?
   ○ Yes
   ○ No

V INDIVIDUAL AND FAMILY CHARACTERISTICS

78. Are you male or female?
   ○ Male
   ○ Female

79. How old were you on your last birthday?

   Age Last Birthday

80. Where were you born?
   ○ In the United States
   ○ Outside the United States to military parents
   ○ Outside the United States to non-military parents

81. Are you an American citizen?
   ○ Yes
   ○ No, resident alien
   ○ No, not a resident alien

82. Did you vote in the last local election? In the last Presidential election?

   A. LAST LOCAL ELECTION

   ○ Yes, in person at the polls
   ○ Yes, by absentee ballot
   ○ No

   B. LAST PRESIDENTIAL ELECTION

   ○ Yes, in person at the polls
   ○ Yes, by absentee ballot
   ○ No

83. Are you of Spanish/Hispanic origin or descent?
   ○ Yes
   ○ No

84. Are you:
   ○ American Indian/Alaskan Native
   ○ Black/Negro/African-American
   ○ Oriental/Asian/Chinese/Japanese/Korean/Filipino/ Pacific Islander
   ○ White/Caucasian
   ○ Other

USE NO. 2 PENCIL ONLY

85. AS OF TODAY, what is the highest school grade or academic degree that you have? DO NOT INCLUDE DEGREES FROM TECHNICAL/TRADE OR VOCATIONAL SCHOOLS. Mark one.

○ Less than 12 years of school (no diploma)
○ GED or other high school equivalency certificate
○ High school diploma
○ Some college, but did not graduate
○ 2-year college degree
○ 4-year college degree (BA/BS)
○ Some graduate school
○ Master's degree (MA/MS)
○ Doctoral degree (PhD/MD/LLB)
○ Other degree not listed above

86. If you are now attending civilian schooling, what kind of school is it? Mark all that apply.

○ Does not apply. I do not attend school
○ Vocational/trade/business or other career training school
○ Junior or community college (2-year)
○ Four-year college or university
○ Graduate/professional school
○ Other

87. What is the highest school grade or academic degree that you think you will complete in the future? Mark one.

○ Does not apply. I don't plan to attend school in the future
○ Less than 12 years of school (no diploma)
○ GED or other high school equivalency certificate
○ High school diploma
○ Some college, but will not graduate
○ 2-year college degree
○ 4-year college degree (BA/BS)
○ Some graduate school
○ Master's degree (MA/MS)
○ Doctoral degree (PhD/MD/LLB)
○ Other degree not listed above

88. Have your parents (or guardians), brothers or sisters (include step-brothers and step-sisters) served in or retired from the military? (Include Guard/Reserve.) Mark all that apply.

	A. Father	B. Mother	C. Brother(s)	D. Sister(s)
Never served	○	○	○	○
Currently serving in the military	○	○	○	○
Served less than 8 years and separated	○	○	○	○
Served more than 8 years (but not retired)	○	○	○	○
Retired from the military	○	○	○	○

89. What is your current marital status? Mark only one answer.

○ Married for the first time
○ Remarried
○ Separated
○ Widowed, GO TO QUESTION 100
○ Divorced, GO TO QUESTION 100
○ Never married, GO TO QUESTION 100

90. Is your spouse currently serving on active duty in the Armed Forces or in the Reserve/Guard?

○ No   ○ Yes, in a Reserve/Guard Component
Yes, on active duty in the:
○ Regular Army   ○ Regular Marine Corps
○ Regular Navy   ○ Regular Air Force
                      ○ Regular Coast Guard

91. Has your current spouse ever served in the U.S. Armed Forces, either on active duty or in the Reserve?

○ No, spouse never served
○ Yes, spouse is retired from Service
○ Yes, spouse is separated from Service
○ Yes, spouse is now in Service

92. How many years have you been married to your current spouse?

Years Married

○ Less than 1 year

○ ○
① ①
② ②
③ ③
④ ④
⑤
⑥
⑦
⑧
⑨

93. How old was your current spouse on her or his last birthday?

Age Last Birthday

○ ○
① ①
② ②
③ ③
④ ④
⑤ ⑤
⑥ ⑥
⑦
⑧
⑨

94. Does your spouse speak English as the main language at home?

○ Yes
○ No

FOR QUESTIONS 95 AND 96 PLEASE MARK THE NUMBER WHICH SHOWS YOUR OPINION ON THE LINE FOLLOWING EACH QUESTION.

How well do you and your spouse agree on your civilian career plans?

Very Well   Not Well At All

①－②－③－④－⑤－⑥－⑦

96. How well do you and your spouse agree on your military career plans?

Very Well   Not Well At All

①－②－③－④－⑤－⑥－⑦

How much of a problem for your family are each of the following? Mark one for each item.

	Serious Problem	Somewhat of a Problem	Slight Problem	Not a Problem	Does Not Apply	Don't Know
a. Absence for weekend drills	○	○	○	○	○	○
b. Absence for Annual Training/ACDUTRA	○	○	○	○	○	○
c. Absence for extra time spent at Guard/Reserve	○	○	○	○	○	○

What is your spouse's overall attitude toward your participation in the Guard/Reserve? Mark one.

○ Very favorable
○ Somewhat favorable
○ Neither favorable nor unfavorable
○ Somewhat unfavorable
○ Very unfavorable

your spouse's support for your decision about staying in the military changed in the past year?

○ Yes, increased
○ No, decreased
○ No, has not changed

EVERYBODY ANSWER:

How many dependents do you have in each age group? Do not include yourself or your spouse. For the purpose of this question, a dependent is anyone related to you by blood, marriage, or adoption, and who depends on you for over half his or her support.

○ Does not apply. I have no dependents.
    GO TO QUESTION 104

NUMBER OF DEPENDENTS

	None	1	2	3	4	5 or More
a. Under 1 year	○	○	○	○	○	○
b. 1 year to under 2 years	○	○	○	○	○	○
2-5 years	○	○	○	○	○	○
6-13 years	○	○	○	○	○	○
e. 14-22 years	○	○	○	○	○	○
f. 23-64 years	○	○	○	○	○	○
g. 65 years or over	○	○	○	○	○	○

101. Are arrangements for your dependents who live with you realistically workable for each of the following situations? Mark one for each item.

○ Does not apply, my dependents do not live with me.

	Yes	Probably	No
a. Short-term emergency situation such as a mobilization exercise	○	○	○
b. Long-term situation such as being called-up or mobilized	○	○	○

102. Are any of your dependents physically, emotionally, or intellectually handicapped requiring specialized treatment or care?

○ No
○ Yes, temporarily
○ Yes, permanently

103. If you are a single-parent or a military member married to a military member, do you have a military family care plan?

○ Does not apply
○ Yes
○ No

104. Do you have elderly relatives for whom you have responsibility even if they are not your legal dependent(s)?

○ No
○ Yes

105. Does this elderly relative live with you?

○ Does not apply
○ Yes
○ No

USE NO. 2 PENCIL ONLY

VI CIVILIAN WORK

A. YOUR OWN EXPERIENCE

106. Are you <u>currently</u>: Mark <u>all</u> that apply.
- ○ Working full-time as an Army or Air Force Guard/Reserve technician, GO TO QUESTION 109
- ○ Working full-time in a civilian job (not technician)
- ○ Working part-time in a civilian job
- ○ With a civilian job but not at work because of <u>temporary</u> illness, vacation, strike, etc.
- ○ Self-employed in own business
- ○ Unpaid worker (volunteer or in family business)
- ○ Unemployed, laid off, looking for work
- ○ Not looking for work but would like to work
- ○ In school
- ○ Retired
- ○ A homemaker
- ○ Other

107. What is your immediate (main) civilian supervisor's overall attitude toward your participation in the Guard/Reserve? Mark one
- ○ Does not apply, I am not working at a civilian job, GO TO QUESTION 109
- ○ Does not apply, I am self-employed
- ○ Very favorable
- ○ Somewhat favorable
- ○ Neither favorable nor unfavorable
- ○ Somewhat unfavorable
- ○ Very unfavorable

108. How much of a problem for your main employer (or for you, if self-employed) are each of the following? Mark one for each item.

	Serious Problem	Somewhat of a Problem	Slight Problem	Not a Problem	Does Not Apply	Don't Know
a. Absence for weekend drills	○	○	○	○	○	○
b. Absence for Annual Training/ACDUTRA	○	○	○	○	○	○
c. Absence for extra time spent at Guard/Reserve	○	○	○	○	○	○
d. Time spent while at civilian work on Guard/Reserve business	○	○	○	○	○	○

THE NEXT QUESTIONS ARE ABOUT YOUR <u>CIVILIAN</u> JOB IN 1991. IF YOU HAD MORE THAN ONE JOB, PLEASE ANSWER THESE QUESTIONS FOR THE JOB WHERE YOU WORKED THE <u>MOST HOURS PER WEEK FOR MOST OF THE YEAR.</u>

109. What kind of work did you do; that is, what is your job called? For example, electrical engineer, construction worker, carpenter, high school teacher, typist, etc.
- ○ I had no civilian job in 1991, GO TO QUESTION 122

WRITE THE NAME OF YOUR JOB IN THE BOX BELOW.

KIND OF WORK/JOB TITLE:

110. Which of the following best describes your civilian employer in 1991? Mark one.
- ○ Federal Government
- ○ State Government
- ○ Local Government (including public schools)
- ○ Self-employed in own business
- ○ Private firm with more than 500 employees
- ○ Private firm with 100-499 employees
- ○ Private firm with less than 100 employees
- ○ Working without pay in family business or farm

111. What kind of organization did you work for in 1991? (For example, TV and radio, manufacturing, retail shoe store, police department, etc. Federal workers: enter the Agency, Department or Government Branch for which you work.)

WRITE THE <u>KIND OF ORGANIZATION</u> (BUSINESS/INDUSTRY) IN THE BOX BELOW. <u>DO NOT</u> WRITE THE NAME OF THE COMPANY.

KIND OF ORGANIZATION:

112. What was your Federal Government pay type and grade at the end of 1991? Mark both the pay type and number grade.
- ○ Does not apply, I didn't work for the Federal Government

A. Pay Type	B. Number Grade	
○ SES or other executive pay	○ 16 or higher	○ 8
○ GM	○ 15	○ 7
○ GS	○ 14	○ 6
○ WS	○ 13	○ 5
○ WL	○ 12	○ 4
○ WG	○ 11	○ 3
○ US Postal Service	○ 10	○ 2
○ Other	○ 9	○ 1

USE NO. 2 PENCIL ONLY

113. In 1991, how many hours per week did you usually work at your (main) civilian job?

Hours Per Week Usually Worked

114. In 1991, how often did you work more than 40 hours per week at your (main) civilian job? Give your best estimate.

○ None   ○ 10-14 weeks
○ 1-4 weeks   ○ 15-19 weeks
○ 5-9 weeks   ○ 20 or more weeks

115. In 1991, how were you paid when you worked over 40 hours a week? Mark one.

○ Not paid extra for working over 40 hours
○ Paid at my regular pay rate for all hours I worked
○ Paid time-and-a-half
○ Paid double time
○ Paid more than double time

116. In 1991, what were your USUAL WEEKLY EARNINGS from your (main) civilian job or your own business before taxes and other deductions? Give your best estimate.

Weekly Earnings

$ | | | | | .00

117. In 1991, how many days of paid vacation did you receive from your (main) civilian job?

Days of Paid Vacation

○ I didn't receive paid vacation

118. In 1991, did you lose opportunities for overtime/extra pay because of your Guard/Reserve obligations?

○ Yes, frequently
○ Yes, occasionally
○ No

119. Which of the following describes how you got time off from your civilian job to meet your Guard/Reserve obligations in 1991? Mark all that apply in each column.

○ Does not apply, I was self-employed, GO TO QUESTION 121

	OBLIGATIONS		
	A. Required Drills	**B.** Annual Training/ ACDUTRA	**C.** Military Schooling
Does not apply, I did not attend	○	○	○
I received military leave/leave of absence	○	○	○
I used vacation days	○	○	○
My Guard/Reserve obligations were on days on which I did not work	○	○	○

120. Which of the following describes how you were paid for the time you took from your civilian job for Guard/Reserve obligations in 1991? Mark all that apply in each column.

	OBLIGATIONS	
	A. Required Drills	**B.** Annual Training/ ACDUTRA
Does not apply, I did not attend	○	○
I received full civilian pay as well as military pay	○	○
I received partial civilian pay as well as military pay	○	○
I received only military pay	○	○
My Guard/Reserve obligations were on days on which I did not work	○	○

USE NO. 2 PENCIL ONLY

121. During 1991, what was the TOTAL AMOUNT THAT YOU EARNED FROM ALL CIVILIAN JOBS or your own business BEFORE taxes and other deductions? Include earnings as a Guard/Reserve technician. Include commissions, tips, or bonuses. Give your best estimate.

Amount Earned at Civilian Job

○ More than $100,000 $ [ ][ ][ ][ ][ ].00

○ None

122. In 1991, how many weeks were you without a job and looking for work?

Weeks Looking for Work

○ I had a job throughout 1991

○ I was not looking for work

123. Do you currently have a spouse?

○ No, GO TO QUESTION 131

○ Yes

○ Yes, separated GO TO QUESTION 131

B. YOUR SPOUSE'S WORK EXPERIENCE

124. Is your spouse: Mark all that apply.

○ Working full-time in Federal civilian job

○ Working full-time in civilian job (not technician or Federal)

○ Working part-time in Federal civilian job

○ Working part-time in civilian job (not Federal)

○ Self-employed in his or her own business

○ With a job, but not at work because of TEMPORARY illness, vacation, strike, etc.

○ Unpaid worker (volunteer or in family business)

○ Unemployed, laid off, or looking for work

○ In school

○ Retired

○ A homemaker

○ Other

125. Is your spouse: Mark all that apply.

○ In the Armed Forces, full-time Active Component, GO TO QUESTION 126

○ In the Armed Forces, full-time Reserve Component (FTS-AGR/TAR), GO TO QUESTION 126

○ Full-time as a Guard/Reserve technician in the Army or the Air Force, GO TO QUESTION 127

○ Part-time in the Guard/Reserve, GO TO QUESTION 127

○ None of the above, GO TO QUESTION 129

126. Was your full-time active duty spouse deployed during Operation Desert Shield/Desert Storm?

○ No, remained at home installation, GO TO QUESTION 129

○ Yes, deployed to the Persian Gulf Area, GO TO QUESTION 128

○ Yes, deployed to other overseas location, GO TO QUESTION 128

127. Was your Guard/Reserve spouse mobilized/activated/called-up for Operation Desert Shield/Desert Storm?

○ No, GO TO QUESTION 129

○ Yes, deployed to the Persian Gulf area

○ Yes, deployed to other overseas location

○ Yes, stayed in our local community

○ Yes, served elsewhere in United States

128. How many months was your spouse on Active Duty during Operation Desert Shield/Desert Storm?

Months

USE NO. 2 PENCIL ONLY

129. In 1991, how many hours per week did YOUR SPOUSE work for pay, either full or part-time, at a civilian job? Give your best estimate.

○ None, GO TO QUESTION 131

Hours Per Week

⓪⓪
①①
②②
③③
④④
⑤⑤
⑥⑥
⑦⑦
⑧⑧
⑨⑨

130. Altogether in 1991, what was the total amount that YOUR SPOUSE earned from a civilian job or his or her own business, BEFORE taxes and other deductions? Include earnings as a Guard/Reserve technician. Include commissions, tips, or bonuses. Give your best estimate.

Amount Earned by Spouse

○ More than $100,000
○ None

$ [   ] .00

⓪⓪⓪⓪⓪
①①①①①
②②②②②
③③③③③
④④④④④
⑤⑤⑤⑤⑤
⑥⑥⑥⑥⑥
⑦⑦⑦⑦⑦
⑧⑧⑧⑧⑧
⑨⑨⑨⑨⑨

VII FAMILY RESOURCES

131. During 1991, did you or your spouse receive any income from the following sources? Mark "YES" or "NO" for each item.

RECEIVED

Yes	No	INCOME SOURCE
○	○	a. Interest and Dividends on Savings
○	○	b. Stocks, Bonds or Other Investments
○	○	c. Alimony, Child Support or Other Regular Contributions from Persons not Living in Your Household
○	○	d. Unemployment Compensation or Workers Compensation
○	○	e. Pensions from Federal, State or Local Government Employment
○	○	f. Pensions from Private Employer or Union
○	○	g. Veterans benefits or pensions
○	○	h. GI Bill
○	○	i. Social Security or Railroad Retirement
○	○	j. Supplemental Security Income
○	○	k. Public Welfare or Assistance
○	○	l. WIC (food programs for women, infants and children)
○	○	m. Government Food Stamps
○	○	n. Anything else not including earnings from wages or salaries

132. During 1991, how much did you or your spouse receive from the income sources listed in Question 131? Do not include earnings from wages or salaries in this question. Give your best estimate.

○ No income from sources in Question 131

○ More than $100,000

$ [   ] .00

⓪⓪⓪⓪⓪
①①①①①
②②②②②
③③③③③
④④④④④
⑤⑤⑤⑤⑤
⑥⑥⑥⑥⑥
⑦⑦⑦⑦⑦
⑧⑧⑧⑧⑧
⑨⑨⑨⑨⑨

133. Overall how do you feel about your/your family income, that is, all the money that comes to you and other members of your family living with you?

○ Very satisfied
○ Satisfied
○ Neither satisfied nor dissatisfied
○ Dissatisfied
○ Very dissatisfied

USE NO. 2 PENCIL ONLY

<u>YOUR RESIDENCE</u>

134. How far is your new principal residence from your last principal residence? Mark one.

○ I have not moved since joining the Guard/Reserve
○ Less than 50 miles
○ 50 to 100 miles
○ 101 to 250 miles
○ 251 to 500 miles
○ More than 500 miles

135. Do you RENT or OWN your principal residence?

○ Neither, live in government-owned or leased housing
○ Neither, live with friends/relatives and PAY NO
    COSTS. GO TO QUESTION 142
○ Neither, live in other accommodations
○ RENT
○ OWN

136. How long have you RENTED or OWNED your residence?

○ 3 months or less	○ 37 to 48 months
○ 4 to 6 months	○ 49 to 59 months
○ 7 to 12 months	○ 5 to 10 years
○ 13 to 24 months	○ 11 to 20 years
○ 25 to 36 months	○ 21 or more years

    If "<u>RENT</u>" continue with Question 137
    If "<u>OWN</u>" go to Question 138

137. How much TOTAL RENT is paid for your residence PER MONTH?

If you share the rent, enter the total rent paid by all occupants. (For example, if it is $525 enter 0525 in the boxes and fill in the matching circles. Include RENT only. Other housing costs will be asked for later.)

Dollars Per Month

$ | | | | | .00

138. What is your monthly house payment for your residence? (Include the PRINCIPAL AND INTEREST on all mortgages or trusts, real estate TAXES and homeowner's INSURANCE. Also include land lease, mobile home lot rental, or berthing fees, if applicable. Other housing costs, such as utility and maintenance costs, etc., will be asked for later. Example: If your payment is $890, enter 0890 in the boxes, then fill in the matching circles.)

Dollars Per Month

$ | | | | | .00

139. Over the last 12 months, what was the AVERAGE MONTHLY cost of all <u>utilities</u> (except telephone and cable TV) <u>paid separately</u> from other rental or home ownership costs?

○ DOES NOT APPLY, No utilities are paid separately
○ Do not have a basis for estimating utility costs

For each utility, add all costs for the LAST 12 MONTHS and divide by 12. (If you do not know the costs for all 12 months, please estimate.)

Enter the average monthly cost for each utility in the space below, then enter the TOTAL at the right.

Dollars Per Month

$ | | | | | .00

	Monthly Average
Electricity	$
Natural Gas/Propane	$
Fuel Oil	$
Wood/Coal	$
Water/Sewer	$
Garbage	$
Total	$

USE NO. 2 PENCIL ONLY

140. Enter the AVERAGE MONTHLY maintenance cost paid for the UPKEEP of the residence. Round off to the nearest dollar.

O No maintenance costs are paid separately

• INCLUDE only maintenance such as plumbing, electrical, heating/cooling system or structural repairs, yard upkeep, etc.
• DO NOT INCLUDE the cost of home improvements (e.g., remodeling, new roof, new furnace, major appliances), new shrubs, new fences, or other additions.
Example: If your cost is $25 per month, enter 025 in the boxes, then fill in the matching circles.

Dollars Per Month

$ [ ][ ][ ].00

141. Enter the AVERAGE MONTHLY cost of any of the following housing expenses for the residence: condominium fee, homeowner's association fee, property and hazard insurance, if NOT included in Question 137 or Question 138.

Fill in the grid for EACH expense you do have or mark "None" for EACH expense you do not have.

	Condominium Fee	Homeowner's Assoc. Fee	Property & Hazard Insurance
	O None	O None	O None

Dollars per Month

Write the numbers in the boxes

Then fill in the matching circles

VIII MILITARY LIFE

How do you feel about the amount of time you spend on each activity listed below? Mark one for each activity.

	I Spend Too Much Time	I Spend About the Right Amount of Time	I Don't Spend Enough Time	Does Not Apply
a. Your civilian job			O	O
b. Family activities			O	O
c. Leisure activities			O	O
d. Guard/Reserve activities			O	O
e. Community activities			O	O

143. The Guard/Reserve are developing new information materials. Below is a list of topics that might be included. How interested would you be in receiving such materials? Please mark your interest in information about each topic.

For each item, mark if you are:	Very Interested	Interested	Somewhat Interested	Not Interested At All
a. Retirement benefits			O	O
b. Survivor Benefit Plan			O	O
c. Family benefits in the Guard/Reserve			O	O
d. Mobilization procedures for dependents			O	O
e. Selected Reserve GI Bill Educational Assistance			O	O
f. Soldiers'/Sailors' Civil Relief			O	O
g. Dental Insurance			O	O
h. Medical Insurance			O	O
i. Mobilization Preparations for Small Business Owners and Partners/Independent Practitioners	O	O	O	O

USE NO. 2 PENCIL ONLY

144. All things considered, please indicate your level of satisfaction or dissatisfaction with each feature of the Guard/Reserve listed below.

For each item, mark if you are:

	Very Satisfied	Satisfied	Neither Satisfied Nor Dissatisfied	Dissatisfied	Very Dissatisfied
a. Military pay and allowances	○	○	○	○	○
b. Commissary privileges	○	○	○	○	○
c. Exchange privileges	○	○	○	○	○
d. Morale/welfare/recreation privileges	○	○	○	○	○
e. Time required at Guard/Reserve activities	○	○	○	○	○
f. Military retirement benefits	○	○	○	○	○
g. Unit social activities	○	○	○	○	○
h. Opportunities for education/training	○	○	○	○	○
i. Opportunity to serve one's country	○	○	○	○	○
j. Acquaintances/friendships	○	○	○	○	○

145. Overall, how satisfied are you with the pay and benefits you receive for the amount of time you spend on Guard/Reserve activities?

Very Dissatisfied ... Very Satisfied

① — ② — ③ — ④ — ⑤ — ⑥ — ⑦

146. Overall, how satisfied are you with your participation in the Guard/Reserve?

Very Dissatisfied ... Very Satisfied

① — ② — ③ — ④ — ⑤ — ⑥ — ⑦

47. We're interested in any comments you'd like to make about Guard/Reserve personnel policies, whether or not the topic was covered in this survey.

DO YOU HAVE ANY COMMENTS?

○ No
○ Yes – Please fill out the COMMENT SHEET on page 23

THANK YOU VERY MUCH FOR ANSWERING THIS SURVEY. PLEASE RETURN IT IN THE ENVELOPE PROVIDED.

COMMENT SHEET

Please provide us with comments you may have regarding Reserve policies or Reserve activities in general in the space below. Before commenting, please fill in one circle in each section.

Your Rank

○ Officer
○ Enlisted

Your Component

○ Army National Guard (ARNG)
○ Army Reserve (USAR)
○ Naval Reserve (USNR)
○ Marine Corps Reserve (USMCR)
○ Air National Guard (ANG)
○ Air Force Reserve (USAFR)
○ Coast Guard Reserve (USCGR)

REFERENCES

Buddin, Richard, and Sheila Nataraj Kirby, *Enlisted Personnel Trends in the Selected Reserve, 1986–1994*, RAND, MR-681/1-OSD, 1995.

Buddin, Richard, and Sheila Nataraj Kirby, *GED Accessions in the Selected Reserve: How Long Do They Serve?* RAND, DB-218-RA (forthcoming).

Burright, Burke, David Grissmer, and Zahava Doering, *A Reenlistment Model for Army National Guardsmen*, RAND, R-2866, 1982.

Department of Defense, *Desirability of Increased Authority to Access Reservists*, Washington, D.C., June 12, 1995.

Grissmer, David W., and Sheila Nataraj Kirby, *Attrition of Nonprior Service Reservists in the Army National Guard and Army Reserve*, RAND, R-3267-RA, 1985.

Grissmer, David W., and Sheila Nataraj Kirby, *Changing Patterns of Nonprior Service Attrition in the Army National Guard and Army Reserve*, RAND, R-3626-RA, 1988.

Grissmer, David W., Zahava Doering, and Jane Sachar, *The Design, Administration, and Evaluation of the 1978 Selected Reserve Reenlistment Bonus Test*, RAND, R-2865, 1982.

Grissmer, David W., Richard Buddin, and Sheila Nataraj Kirby, *Improving Reserve Compensation: A Review of Current Compensation and Related Personnel and Training Readiness Issues*, RAND, R-3707-FMP/RA, 1989.

Grissmer, David W., Sheila N. Kirby, Man-bing Sze, *Factors Affecting Reenlistment of Reservists: Spouse and Employer Attitudes and Perceived Unit Environment*, RAND, R-4011-RA, 1992.

Grissmer, David W., Sheila N. Kirby, Man-bing Sze, *Insuring Mobilized Reservists Against Economic Losses: An Overview*, RAND, MR-446-OSFD, 1995.

Kalton, G., *Introduction to Survey Sampling*, Sage Publications, Beverly Hills and London, 1983.

Kirby, Sheila N., and David W. Grissmer, *Reassessing Reserve Attrition: A Total Force Perspective*, RAND, N-3521-RA, 1993.

Marquis, M. Susan, and Sheila Nataraj Kirby, *Prior Service Individuals in the Army National Guard and Army Reserve*, RAND, R-3686-RA, 1989.

Perry, William, Memorandum for Secretaries of the Military Departments on *Increased Use of Reserve Forces in Total Force Missions*, Department of Defense, Washington, D.C., 7 April 1995.

Rizzo, Louis, David Morganstein, Veronica Nieva, and Shelley Perry, *Weighting Report for the 1992 DoD Reserve Components Surveys of Officers and Enlisted Personnel and Their Spouses*, WESTAT, Inc., Washington, D.C., 1995.